Living with the South Carolina Coast

Gered Lennon, William J. Neal, David M. Bush, Orrin H. Pilkey,

Matthew Stutz, and Jane Bullock

Duke University Press   Durham and London   1996

The Living with the Shore series is funded by the Federal Emergency Management Agency

Living with the Shore Series

Publication of 19 volumes in the Living with the Shore
series has been funded by the Federal Emergency Management Agency.

Publication has been greatly assisted by the following indi-
viduals and organizations: the American Conservation As-
sociation, an anonymous Texas foundation, the Charleston
Natural History Society, the Office of Coastal Zone Man-
agement (NOAA), the Geraldine R. Dodge Foundation, the
William H. Donner Foundation, Inc., the George Gund
Foundation, the Mobil Oil Corporation, Elizabeth
O'Connor, the Sapelo Island Research Foundation, the Sea
Grant programs in New Jersey, North Carolina, Florida,
Mississippi/Alabama, and New York, The Fund for New
Jersey, M. Harvey Weil, Patrick H. Welder, Jr., and the
Water Resources Institute of Grand Valley State University.
The Living with the Shore series is a product of the Duke
University Program for the Study of Developed Shorelines,
which was initially funded by the Donner Foundation.

© 1996 Duke University Press

Printed in the United States of America

on acid-free paper ∞

Library of Congress Cataloging-in-Publication Data

appear on the last printed page of this book.

Orrin H. Pilkey, Sr., 1905–1994

The Living with the Shore series was
inspired by my father's experiences
with Hurricane Camille (1969).
This, the nineteenth volume in the series,
is dedicated to his memory.

Orrin H. Pilkey

Contents

Figures, Tables, and Risk Maps

Risk Maps

Note: Original risk maps have been reduced to accommodate the scale of this book.

Preface

When we set out to produce a second edition of *Living with the South Carolina Shore,* we found that a simple rewrite wouldn't suffice. Nothing stays the same, as they say, and the coast had changed since 1980, rearranged by Hurricane Hugo (1989) and several lesser storms; our knowledge of barrier islands and our approach to evaluating risk to coastal development had changed, expanded as a result of looking at other shorelines and our continuing focus on improving island-wide mapping and risk mitigation; construction trends had changed, with beach "cottages" replaced by beach "mansions" or multiple-dwelling structures in the post-Hugo recovery; and South Carolina's approach to coastal zone management had changed, legislated to work more closely with nature and to reduce future property losses. We also found a few hazards that we had previously overlooked.

Some of the original material remains because some things haven't changed. Nature still works the same way. Development pressure in the coastal zone continues, especially on the high-risk barrier islands. People still want to locate their houses in flood zones, out of ignorance or, in the case of some post-Hugo examples, perhaps arrogance. The taxpayer-at-large is still picking up the tab for too many "donuts" in each disaster; only the "donut holes" now equal billions of dollars rather than millions! People still remove maritime forest, level dunes, block overwash, and otherwise interfere with the natural system and con-

tribute to worsening the impact of coastal hazards. Generations of previous jetties, groins, and seawalls (25 percent of South Carolina's developed shore is "stabilized") still contribute to downdrift erosion problems. People still ask, "When is *someone* is going to do something about *my* eroding beach?" And barrier island problems still figure prominently in the daily media reports and politics at all levels. Sometimes we are so focused on a current issue—such as to relocate or not relocate a structure, or to replenish or not replenish a beach—that we overlook future humbling hazards such as flooding or earthquakes and their associated problems. In order to address the numerous changes in South Carolina, the specific lessons learned from Hurricane Hugo, and the natural processes and potential issues that often stay the same, we produced a mix of new book and revised old book. Thus, instead of a second edition, we present a new book with a new title: *Living with the South Carolina Coast.* The term *coast* denotes a broader zone than *shore.* Technically, the shore is the line where water meets land. The coast is the entire offshore and onshore area influenced or affected by the meeting of land and ocean water. *Coast,* we feel, more accurately reflects the scope of this book.

Most significant, we have been joined by new authors who bring an up-to-date view of the South Carolina coast and a breadth of experience in coastal hazard evaluation and mitigation. Gered Lennon's master's thesis maps

the faults responsible for the 1886 Charleston earthquake and its related effects—a hazard that few coastal residents ever consider. Gered is a former staff member of the South Carolina Coastal Council and is thoroughly familiar with the state's coastal zone management policies, problems, and politics. He served with the South Carolina Coastal Council during the time the Beachfront Management Act was implemented and during the preparation for, and recovery from, Hurricane Hugo. More important, Gered lives where and what he writes—in the coastal zone. As a resident of Folly Beach, he knows firsthand how a community with a serious erosion problem is torn between the power of nature, the impacts of hazards, trying to mitigate those impacts, and still having a viable community of people drawn by the beauty and spirit of the sea and its shores.

David Bush of West Georgia College in Carrollton, Georgia, was a resident of North Carolina for many years and is a shoreline expert. While he was at Duke University his research focused on coastal hazards, risk assessment mapping, and property damage mitigation. He knows much of the Atlantic seaboard and the Caribbean well, especially the Carolinas and Puerto Rico. His expertise was recognized in his appointment to the National Academy of Sciences Post-Disaster Field Study Teams after Hurricanes Gilbert and Hugo. He was involved with planning efforts for the U.S. Decade for Natural Hazard Reduction and is the senior author of *Living with the Puerto Rico Shore* and *Living by the Rules of the Sea*.

The late Orrin Pilkey, Sr.'s, revised construction chapter includes new concepts and design recommendations based on the lessons taught by Hurricanes Andrew and Iniki. Jane Bullock, senior policy adviser to the director of the Federal Emergency Management Agency, wrote the earthquake construction chapter. In her long tenure with FEMA, Jane has dealt with numerous hurricanes, earthquakes, and flood disasters.

Matthew Stutz received his first taste of coastal hazard assessment, mapping, and mitigation when he spent the summer and fall of 1994 as an undergraduate research intern at Duke University studying with Orrin Pilkey and David Bush. Matt did much of the fieldwork, making and checking the first round of risk maps. He is currently pursuing his graduate studies on a Fulbright Scholarship.

In the preface to *Living with the South Carolina Shore,* which is reprinted below, we acknowledge those who have supported the series, but we must recognize an additional cast of individuals without whom this book would not have been possible. One of the driving forces behind this book and the entire Living with the Shore series has been the support of the Federal Emergency Management Agency. Specifically, the FEMA officials who provided input for this latest volume are John McShane, Gary Johnson, Jane Bullock, Mark Russo, and Dick Krimm. The conclusions of this book, however, are those of its authors, based on published reports and studies of record, and are not meant to reflect the views of any specific agency.

Amber Taylor drafted most of the figures for the book and oversaw the drafting of the risk maps and line drawings. Thanks to Debbie Gooch for typing several drafts. Craig Webb, Rob Thieler, Rob Young, Susan Bates, Art Trembanis, and Carmen Bush helped with many tasks along the way. Thanks also to John Cooler of Beaufort, pilot extraordinaire, and photographer Victoria Massina of Folly Beach for their assistance with aerial photography.

William J. Neal
Orrin H. Pilkey
Series Editors
August 1, 1995

Preface 1984

Just over a decade ago my family's favorite holiday was to visit South Carolina's sea islands. We'd pack a picnic, load the kids in the car, and set off from our home in Statesboro, Georgia, to our favorite beach on Hilton Head Island. As we wound our way down U.S. 278, everyone's anticipation grew, and at the first sight of salt marsh we knew we were in sea island country! We parked on a public lot of crumbling asphalt, separated from the beach by a thicket into which several well-worn paths led—a natural changing room from which we emerged onto the long, wide beach. In late winter and early spring we could feel isolated with nature, play chase from beach to thicket, collect shells, build sand castles, or wade at the surf's edge. If you walked far enough along the beach in either direction there was development, but it was often back from the beach and fronted by vegetation, seemingly removed from this bit of Eden.

We were not aware that the shoreline was eroding. True, a shrub might have toppled here or there along a small scarp at the back of the beach, and there was the black sand concentrate from the last storm's swash, washed into the edge of the thicket. An erosion "problem" did not exist. The beach was the seaward side of a barrier island, moving as the island moved through time, together with marshes, dunes, thickets, tidal creeks, and inlets. Still, we had no feeling for this movement, for we were like the other sunbathers, beachcombers, swimmers, fishermen, and recreationists—concerned only that there was a beach rather than where the beach was. However, a few hundred yards to the north something different from the thicket or dune's toe was in the path of this moving shore. Just how different would be apparent to us 10 years later.

Shoreline changes often are spectacular only if seen over a time longer than a few years. In 1981 we returned to our favorite beach hoping to catch a glimpse of the happy past. The parking lot was newly paved. The thicket was essentially gone. And one certainly wouldn't consider changing to beachwear behind a bush with a Holiday Inn flanking one side and a condominium the other. Down on the beach among the shells we found angular rock chips (ouch!) of granite, a rock very much out of place in South Carolina's coastal plain. With a bit of detective work we noted the chips increased to the north, now giving way to a chunk-size piece here and there. The source was soon apparent—a seawall. A row of cottages now sat in jeopardy at the back of the beach, caught between their property lines and a collapsing sea scarp. The residents had made a hard choice, a choice already costly to them as well as to the public. This beach as we knew it will be destroyed. Many of the people who once came here will now go elsewhere. Unfortunately, the situation I've just described is not an isolated one. If you've been coming to the South Carolina coast regularly, or if you've resided here even for a relatively short time, you can tell similar stories about threatened or destroyed property: the walls at North Myrtle Beach, Folly Island, or Seabrook; the various groin fields and the beach nourishment projects up and down the coast. The dilemma of the South Carolina coastal property owner is growing, and the purpose of this book is to address the underlying causes and suggest some solutions. The intent is not to discourage development but rather to encourage proper *limited* development.

The following text is primarily directed to the coastal dweller and the coastal developer, the people who have built or bought on the shoreline or anywhere on a barrier island, and those who plan to do so. It should also be valuable to real estate agents, bank loan officers, contractors, investors, and others dealing with the coast. Our goal is to increase the reader's awareness of how barrier islands and beaches operate, how shorelines and entire barrier islands may retreat, what kinds of hazards are faced by coastal dwellers and property owners, and how to reduce the impacts of those hazards if you are already in such a zone. Perhaps your choice will be more prudent than that of the multitude whose property is threatened by such hazards.

The information presented here represents the summation of the work of many investigators, too numerous to mention, but to whom we owe a sincere thanks. Their respective agencies include the Coastal Research Division, Department of Geology, University of South Carolina; the South Carolina Wildlife and Marine Resources Department; the South Carolina Coastal Council; the South Carolina Sea Grant Program; and the U.S. Army Corps of Engineers, Charleston District.

This book is one of a series of citizen's guides to barrier islands being produced under the direction of Dr. Orrin H. Pilkey, Jr., Duke University, Durham, North Carolina, and Dr. William J. Neal, Grand Valley State Colleges, Allendale, Michigan, and published by Duke University Press. Their work has been supported by grants from the Coastal Zone Management Program of the National Oceanic and Atmospheric Administration. This program is administered through the University of North Carolina Sea Grant Program, and more recently by the Federal Emergency Management Agency. The efforts of W. Carlyle Blakeney, Jr., Vice President, National Audubon Society, Charleston, South Carolina, were made possible by a Carl W. Buchheister Fellowship awarded by the National Audubon Society. The authors acknowledge this support with sincere thanks. The conclusions reached herein, however, are those of the authors based on published reports, maps, air photos, and field observations, and do not reflect the official policies or viewpoints of the supporting agencies. The chapter on hurricane-resistant construction was authored by Orrin H. Pilkey, Sr., a retired civil engineer, of Waveland, Mississippi, who has had firsthand experience with hurricanes, particularly Hurricane Camille in 1969. Special thanks are extended to Bette Weerstra for typing and retyping most of the manuscript; Clint Myers and Barbara Gruver for drawing the line illustrations; and Martin Wilcox and Doris Schroeder for their editorial contributions.

As in any effort of this sort, we were helped by many people who live and work along the shore—again, so many we cannot list them all. We are grateful for their cooperation and for the insights and concerns we acquired from them. We dedicate this work to them and all of those who come to enjoy South Carolina's coast.

William J. Neal

On the evening of September 21, 1989, Americans helplessly watched news reports of Hurricane Hugo's swath of destruction across the Caribbean. To many, it was simply one more scene of destruction from somewhere else in the world. But on this night the residents of South Carolina were about to be linked to the same reality that had devastated portions of the West Indies, including the Virgin Islands and Puerto Rico. Very few imagined the future in store for them. Tom Brokaw's NBC News sign-off was an ominous message for viewers in South Carolina: "As we sign off, things do not look good for Charleston."

It was not to be just Charleston, however. In the South Carolina coastal zone 900,000 people and billions of dollars worth of property were now in the path of one of the earth's greatest forces. The most vulnerable area included the state's fringe of fragile barrier islands (fig. 1.1) and the concentration of people and property located on them. Fortunately, thousands had started evacuating inland on the evening of September 20, more than 24 hours before Hugo struck. By the time Hugo made landfall, more than 186,000 persons had left their homes, doubtless saving many lives. How did these people come to be in such a dangerous area, and why was so much capital invested in the property now at risk? A more important question is: How has the Hugo experience changed the way our coastal zone is managed?

All the amenities that people seek along the shore are available on the South Carolina coast. The northern coast, known as the Grand Strand, offers 30 or more miles of sandy beaches. Myrtle Beach is the classic family-oriented amusement and beach complex. At the southern end of the state, Hilton Head Island has a proud image as an exclusive resort and is known internationally for its golf and tennis tournaments. In between, Kiawah, Seabrook, and Fripp Islands are developments cast in the image of Hilton Head. Pawleys Island's fame comes from its namesake hammock, which symbolizes the casual island liv-

1.1. Index map of the South Carolina coast.

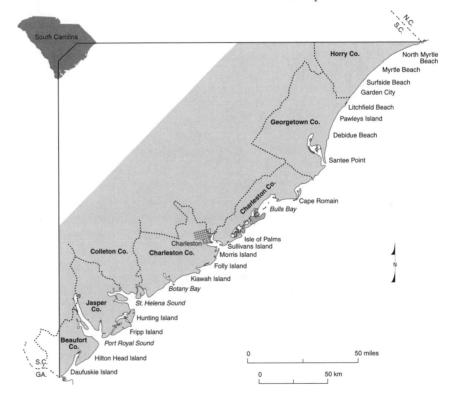

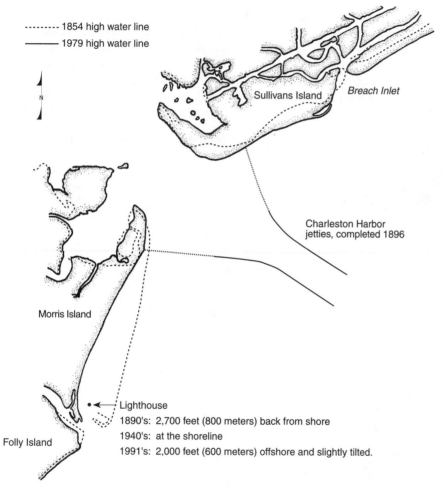

- - - - - - - 1854 high water line
————— 1979 high water line

Sullivans Island   *Breach Inlet*

Charleston Harbor
jetties, completed 1896

Morris Island

Lighthouse
1890's: 2,700 feet (800 meters) back from shore
1940's: at the shoreline
1991's: 2,000 feet (600 meters) offshore and slightly tilted.

Folly Island

1.2. Map of the entrance to Charleston Harbor showing the jetties that were completed about 100 years ago. Largely as a result of the jetties, Morris Island, the site of the events chronicled in the movie *Glory*, has migrated more than 4,700 feet. The lighthouse, once on the back side of the island, is now 2,000 feet offshore. *Source:* Adapted from *An Illustrated History of Tidal Inlet Changes in South Carolina* (appendix D, ref. 68).

ing enjoyed there since the days of the rice planters 200 years ago. Edisto Beach is a more recent counterpart. The Isle of Palms, Sullivans Island, and Folly Beach, all suburbs of Charleston, host a colorful mix of islanders, commuters, and beach enthusiasts. Other islands, including parts of Edisto Beach and Hunting Island, are state parks where the public may enjoy a natural beach environment. A few areas, like much of the Santee Delta and the estuarine ecosystem along the Ashepoo, Combahee, and Edisto Rivers (the ACE Basin), are wildlife preserves. A large area of the ACE Basin is now protected through a coalition of private plantation owners working in cooperation with conservation groups.

A peak summer day finds 350,000 people in the Grand Strand area—more than 10 times the number of year-round residents—35,000 on Hilton Head, and tens of thousands more enjoying the islands in between. Each year, more and more of these visitors decide to remain on the coast as permanent or part-time residents. In fact, the development rate of South Carolina's coastal zone is comparable to an old-fashioned boomtown. Horry County, with a population growth of 42 percent from 1980 to 1990 (appendix D, ref. 7), is a prime example: two-thirds of the county's 150,000 residents live between the Atlantic Ocean and the Intracoastal Waterway, and own more than $1.4 billion worth of oceanfront real estate in the city of Myrtle Beach. Along with the permanent residents and visitors has come a concomitant increase in hotels, condominiums, golf courses, and other amenities.

Unfortunately, enthusiasm for the beach is often accompanied by loss of awareness of the basic nature of the coast and the processes that take place there. To illustrate this fact, consider that most of the people affected by Hurricane Hugo had never experienced a severe coastal storm of any kind. Yet a realistic, longer-term view shows that these storms strike with statistical regularity and frequency. Not only do they shape the shoreline, they also help to define the nature of life along the coast.

Barrier Islands

South Carolina is blessed with particularly beautiful and diverse barrier islands. Long, thin islands parallel the shore along much of the coast. Barrier islands are dynamic; that is, they react to coastal and oceanic conditions. For instance, they are constantly altered by storms, while at the same time they protect the mainland by acting as natural shock absorbers against storm forces. They migrate landward, changing position to avoid being drowned as the world's sea level rises. Barrier islands also react to the conditions we humans create, especially as we increasingly develop the coast and attempt to prevent the effects of storms and rising sea level. Consider, for example, the case of Morris Island.

An 1848 report (appendix D, ref. 65) describes Morris Island as follows:

On the beach side, [there is] a long line of sand-hills, rising to the height of thirty or forty feet, and, in some places, covered with pines and palmettoes. . . . The hills, towards the middle of the island, rest upon a mud flat. As they are driven backwards by the winds, or washed away in front by the waves, the flat becomes exposed. For more than a mile in length there are numerous stumps in this flat, which, on examination, prove to be tap roots of pines, which had penetrated through the sand upon which they once stood, and which has been washed away, leaving the stumps, as if they had grown on the mud flat, which is now below high-tide.

In the mid-nineteenth century, Morris Island was a healthy, migrating barrier island with dune ridges and forest. Fifteen years later, the island was the scene of a ferocious struggle between the Blue and Gray armies in the battle for Fort Wagner and control of Charleston Harbor. A century after that it was little more than a flat strip of land frequently overwashed by waves.

The brick lighthouse on Morris Island has been a well-known coastal landmark since 1874, when it was constructed 1,300 feet inland from the lighthouse it replaced, a casualty of the Civil War. The immobile structure is a reference point against which shoreline changes can be measured (fig. 1.2). In the late nineteenth century it sat approximately 2,700

1.3. Morris Island lighthouse going out to sea. Photograph taken by George W. Johnson around 1940 to 1950; from the collection of W. J. Keith.

1.4. Morris Island lighthouse today. The island in the background is Folly Island; Morris Island is out of the picture to the right.

feet back from the shoreline. By 1940 the lighthouse was at the shoreline (fig. 1.3). Today the lighthouse stands, slightly tilted, approximately 2,000 feet offshore (fig. 1.4). It marks 4,700 feet of shoreline retreat.

Most barrier islands, and most coastlines in general, are retreating. Sea levels are rising worldwide, and the islands respond by migrating—moving toward the mainland. But Morris Island has retreated more than 4,700 feet in just under 100 years, a rapid retreat that reflects another situation to which the island has also had to respond: the jetties constructed at the entrance to Charleston Harbor. Doubtless Morris Island would have been no stranger to coastal erosion had the jetties never been built; without the jetties' influence, however, the island today would be more like it was in 1848.

The jetties, two massive walls 2.5 and 3 miles long extending out from Sullivans Island and Morris Island, were completed in 1896. Their purpose is to keep sand from building up in the harbor entrance and interfering with ships moving in and out of Charleston Harbor. Unfortunately, the jetties also cut off the sediment supply to the shallow sand bodies in front of Morris Island that once formed the shoals of the immense delta of Charleston Harbor. As these protective sand bodies disappeared, Morris Island was exposed to greater wave attack, and it also received less incoming sand to replace sand that was lost. Thus, an island once characterized by forested dune

ridges and large enough to be occupied by thousands of troops during the War Between the States has become a low sand flat.

The effect of the Charleston jetties on Morris Island is not an isolated case. The structures that we build to protect some particular part of the coastal environment often end up doing damage to it. In some places so many structures have been built that the damage has been catastrophic. We need only look at the experience of New Jersey for confirmation.

New Jerseyization: The End of the Beach

The well-known environmental crises and property losses that occurred on the New Jersey shoreline—called "New Jerseyization" by coastal geologists, engineers, and planners—occurred when residents, planners, and developers failed to recognize the basic natural processes of the shoreline (fig. 1.5). Those losses should help to guide today's developers away from similar mistakes.

New Jersey's shoreline development began around 1800 when access to the shore improved, then proceeded very rapidly because of the proximity of large population centers. Hotels and cottages were crowded together on every available piece of land, with little thought to the safety of the sites. Often these structures were not as well built as inland dwellings, despite the fact that they would have to endure

1.5. New Jerseyization. (A) The real thing in Seabright, New Jersey. The seawall blocks the view of a beachless seashore and protects property worth less than the value of the wall. (B) On the road to New Jerseyization on Fripp Island, S.C.

far greater natural forces. The new buildings were soon threatened by natural coastal processes. People were concerned that the beaches seemed to be eroding away. In response, New

Jersey armored its shoreline, hoping to protect the beaches and development by building seawalls, bulkheads, and groins.

Today the remains of many of those protection schemes clutter the shore. In some places the beach has completely disappeared. A trip to the New Jersey shore would be worthwhile for every South Carolinian: one look can convey a much more dramatic message than the pages of any book.

New Jerseyization is not only destroyed beauty (after all, some people prefer to see a row of condominiums on the beach rather than a line of dunes covered with sea oats), it is also a serious threat to coastal residents. Hurricanes, increasing costs, pollution, environmental destruction, and reduced public access all threaten coastal residents, in very different ways.

Hurricanes

Where there is the constant threat of hurricanes, lives are endangered by hazardous building sites and unsafe construction. Unfortunately, development on the South Carolina coast has been attracted to the higher-risk sites. Too often the most eager developers have owned the most dangerous parts of the islands. Poor-quality construction is another danger, independent of the safety of the building site. Hurricane Hugo's multi-billion-dollar price tag clearly underlines this point.

Increasing Costs

Even though new seawalls are prohibited on the shores of South Carolina, maintenance and repair of the existing walls can be very expensive. Any type of shoreline engineering initiates a spiraling increase in costs, both in development landward of the beach and in the real expenditures for future erosion control. Shoreline engineering, whether it is a seawall or beach nourishment, fosters a false sense of security in the community, evidenced by larger structures built, where possible, even closer to the water. The greater investment in turn increases the cost-to-benefit ratio used to evaluate the engineered beach. In the South Carolina experience, any storm protection provided by renourishment often disappears in the first year or so. Storms smaller than Hugo have demonstrated the vulnerability of seawalls. The community's poststorm attention focuses on its failed or destroyed "shoreline protection," and citizens resolve to rebuild the beach or seawall "bigger and better," still trying to oppose a force that is unstoppable. Beachside mansions provide the "justification." Everybody pays, and then pays more, all the while unwittingly moving in the direction of New Jerseyization.

Pollution

Improper waste disposal threatens the health of coastal citizens and destroys the natural resources that support the local marine fishery. Whether through storm or other mishap, broken or failed sewage systems have forced the closure of South Carolina shellfish grounds numerous times.

Environmental Destruction

The beach—the very environment we rush to the shore to enjoy—is ultimately destroyed when it is overdeveloped. Scenic dunes, maritime forests, and marsh habitats disappear. This alteration of the environment is the most striking aspect of New Jerseyization. Beach-saving devices work only temporarily at best. Where seawalls are built, the beach is eventually lost. Old beach resorts in New Jersey and Florida have no beaches at all except where sand has been pumped in. In addition, beach repair is very expensive to the taxpayer. The estimated cost to renourish South Carolina's beaches in the 1990s alone is $65 million. Under the Beachfront Management Act, South Carolina has embraced beach nourishment as the means to cope with eroding shorelines where "economically justified."

Reduced Public Access

Private development inevitably reduces access to the beach for the public, who nonetheless must often pay the bills for beach repairs. Access to the beach is frequently prohibited to all

1.6. Construction on Pawleys Island before and after Hugo. (A) Pre-Hugo houses on the southwestern end of Pawleys Island. (B) Post-Hugo replacement houses in the same area are bigger but no better with respect to potential damage from future hurricanes.

but adjacent property owners; others must pay access charges. South Carolina has taken more than just a few steps down the road to New Jerseyization. The high-rise buildings con-structed virtually on the beach, missing sand dunes, houses at the high-water mark, and various "protective" devices are but a few signs of the trend. But there are encouraging signs as well. In 1988, the Beachfront Management Act was added to the Coastal Zone Management Act (see chapter 10). The new law greatly increases the state's power to prevent unwise development of beach areas. The South Carolina Department of Health and Environmental Control's (DHEC) Office of Ocean and Coastal Resource Management (OCRM), formerly the South Carolina Coastal Council, is mandated by this law to monitor the condition of the beaches and use the data it collects to establish setback areas. Building in erosional areas is prohibited. Perhaps more important in the long run, the new law prohibits the building of any new shore-hardening structures on South Carolina's beaches.

An Alternative to New Jerseyization

Development will continue on the South Carolina coast. Since Hurricane Hugo, rebuilding and development have tended toward "more and bigger," in some cases even closer to the beach (fig. 1.6), in spite of the Beachfront Management Act. To expect this process to stop is unrealistic; however, development must proceed intelligently if we are to avoid the mistakes that have been made here and elsewhere.

The purpose of this book is to provide South Carolinians with the information they need to make intelligent decisions about coastal development. Thus, after closing this chapter with a brief history of development and storms along the South Carolina coast, we will outline the natural processes that are at work at the shore (chapter 2), focusing on Hurricane Hugo and similar events (chapter 3) and describing the various ways people have tried to control these processes (chapter 4). Chapter 5 is a primer on assessing risk, and chapter 6 contains a site-by-site description of the South Carolina coast, describes the natural hazards present at each, and gives information about how to evaluate possible development sites. Chapter 7 discusses hazards of the coastal zone other than storms (earthquakes, fires, and floods). Guidelines for building in a coastal environment are given in chapter 8, and chapter 9 offers suggestions for building earthquake-resistant structures. Finally, chapter 10 outlines the laws that apply to land use in the South Carolina coastal zone. These discussions are supplemented by four appendixes: (A) a set of natural hazards safety checklists; (B) a guide to agencies involved in coastal development, regulation, and information; (C) the Beachfront Management Act; and (D) an annotated list of useful references.

The Foundations of Development: Historical Perspective

The First Settlements

The Spanish sailed into Winyah Bay as early as 1521, and established the first European settlement in South Carolina there in 1526. Cold, fever, starvation, and poor relations with the natives caused the settlement to fail. Thirty-six years later, French Huguenots established Charlesfort on Parris Island, only to fail as the Spanish had before them. In turn, the Spanish established forts on Parris Island between 1566 and 1587, but these were military outposts, not attempts to colonize the New World.

By the seventeenth century the English laid claim to all of "these lands." In 1665 King Charles II granted the territory to eight lord proprietors, opening the way for English settlements. The proprietors established the initial land divisions, but later, in 1729, they sold their rights back to the Crown, and the king made new land grants to individuals. Modern landownership in South Carolina still traces back to these original grants, occasionally creating disputes. Some of the seashore land of the 1600s is now under the Atlantic Ocean, and some of the dry land of those days is now regulated or protected wetland.

Charles Towne, established in 1670 at Charles Towne landing on the banks of the Ashley River, became the first successful settlement. It was moved to its present site in 1680 (and renamed Charleston in 1783). Beaufort was established in 1710, and George Town in 1729. These first towns were all seaports, and they were located in protected areas in the estuaries.

The coastal land parcels granted by the Crown extended from inland to the sea. Percival Pawley, for example, was given 13 grants, one of which extended from the Waccamaw River to the "sea marsh." Thus, a pattern of long, narrow plantations was established. The plantation beachheads were the ancestral summer places that became today's developments. The lowlands, wetlands, and navigable streams included within their boundaries were important agriculturally, first for rice, later for indigo, and finally for cotton. The environmental characteristics of the low country and the sea islands were particularly important to the tidal method of rice cultivation. By developing a system of dikes and canals, rice growers took advantage of tidal reaches of coastal rivers and creeks to flood the rice fields. The water routes also provided access to the barrier islands—which would continue to be developed long after hurricanes had destroyed the rice-field dikes and had contributed to the demise of rice cultivation.

Some barrier islands were the outer fringes of plantations, but others were central to plantation activity. John Stanyarne built a four-story mansion on Kiawah Island in 1772 (the Vanderhorst house, pronounced "Vandrost"), and the 1790 census of the island recorded five whites and 52 black slaves. The island had previously supported a small Indian population. Judging from the accounts of early hurricanes, there were slave populations on several of the islands during summer and fall.

Plantation agriculture was not the only activity on the barrier islands in the eighteenth and nineteenth centuries. Morris Island was the site of temporary lights used to guide ships into Charles Towne harbor as early as 1673, and a permanent lighthouse was constructed there in 1767. The present Morris Island lighthouse was first lit in 1876. Other lighthouses include the first and second Cape Romain lighthouses (1827, 1857) on Lighthouse Island, and the Hunting Island lighthouse, built in 1875 to replace one erected in 1859. The Hunting Island lighthouse was moved to its present position in 1889 after it was threatened by coastal erosion. Lighthouses were necessary then for safe travel and commerce in coastal waters, but their use has declined with the advent of satellite navigation. The few still in use are testimony to a once-proud tradition of the sea (see appendix D, ref. 6, for a detailed history of lighthouses).

Quarantine hospitals (also called pesthouses or lazarettos) were often set up on barrier islands adjacent to harbor entrances to prevent victims of cholera, yellow fever, malaria, and similar "fevers" from coming ashore. One such facility was constructed near Fort Johnson in 1796 to replace an earlier pesthouse on Sullivans Island. Morris Island and Folly Island were similarly used in the 1830s. Ironically, while ships were lying at anchor to discharge their patients, healthy passengers were being infected by mosquitoes coming from the hospitals! The remains of the Fort Johnson facility were destroyed in an 1885 hurricane.

One of the more innovative coastal projects of the day was based on the use of wind and solar energy. A saltworks developed on Pawleys Island in 1782 used a windmill to pump seawater into evaporating vats to recover the salt. The product was important to the American Revolution, and production continued into the War Between the States. Union troops destroyed the works and a similar operation at Singleton Swash in the North Myrtle Beach area. Now, more than 100 years later, we are rediscovering Atlantic breezes and coastal sunshine as energy sources.

Most of the people who came to the barrier islands were there to take part in military operations. Succeeding the earlier Spanish, French, and British fortifications along the coast, the Americans established Fort Moultrie on Sullivans Island in 1776, the famous palmetto fort for which South Carolina is nicknamed. Now a national park, the fort was in discontinuous use from the Revolution through World War II. British troops used Morris Island in 1780 and invaded Seabrook (Simons) Island the same year. The most intense military activity on the South Carolina

coast, however, occurred during the War Between the States. Thousands of Union troops were stationed on Folly Island, from which they launched the fierce battle for Morris Island and Fort Wagner that lasted from July 10 to September 7, 1863 (see appendix D, ref. 8, for a description of the battle). Forts Moultrie and Sumter remained Confederate strongholds throughout most of the war. Other islands saw more limited action with small gun emplacements (e.g., Hilton Head Island) or temporary troop invasions (e.g., Seabrook Island).

Lighthouse keepers, slaves, fever victims, and soldiers were on the islands by necessity rather than choice. Their thoughts were on survival, not recreation. Some died from fever, some died from wounds, and many died at the hand of nature. Some thought they would die of boredom or loneliness while waiting to escape to the mainland. Now, we seek to reverse the process, to escape from the mainland to the barrier islands, where we are at leisure to enjoy the amenities of the sea.

The First Resorts

Although they were not recognized as carriers in the 1700s, mosquitoes transmitted a number of maladies that went by such names as "summer fever," "country fever," and "night miasmas." Plantation families discovered that the coast was a good place to escape from these fevers. From early May to November the entire family moved to the vicinity of the plan-

tation beachhead. Pawleys Island (Allston's Bank) was used for recreation by 1768 and had summer houses by 1790. When the Marquis de Lafayette first landed in America, on North Island, he found hospitable inhabitants there. A village grew at North Inlet to include 100 houses and a church. Although it was swept away in the hurricane of 1822, the village was rebuilt, complete with a school. Homes were built on the Santee River Delta islands (Murphy Island, Cedar Island, and South Island). Cedar Island had a small village. Other summer communities developed at Murrell's Inlet and on Debordieu Island (also known as Sandy Island, Dubourdieu Island, Dubordia Island, and Debidue Beach). Surprisingly, the Myrtle Beach area was not the site of early development, although George Washington visited residents in the vicinity of Windy Hill Beach and Singletons Swash in 1791 (appendix D, ref. 1). These early residents had tastes like our own. They enjoyed sea bathing, beach walks, shelling, taking the air, horseback riding, and good food aplenty from fishing, shrimping, and crabbing. We can speculate that they would also have enjoyed dance pavilions, arcades, and water slides.

Pawleys Island in the 1800s is a good example of the long-term progress of development. From English land grant, to property divisions of a few hundred acres, to 5- to 10-acre plots, to smaller and smaller lots, the island's history is one of subdivision, culminating in the smallest of modern units—the condominium.

By the 1890s seaside resorts were in vogue. Inns and boarding houses were built on Pawleys Island, along with a barracks for dancing. Myrtle Beach began as a resort for visitors rather than local families. The village was named in 1900, and the first hotel opened in 1901. Its early growth was slow, but Myrtle Beach hasn't stopped growing since then.

Smaller resorts followed. Folly Beach opened in 1918, at first as a summer tent city. By the early 1930s a post office, the first apartment house, and, later, a dance pavilion gave the town a permanent look. Other islands and coastal stretches were developing in the 1930s (e.g., Edisto Island, Seabrook Island, and Atlantic Beach). Although the first resorts may have been no more than church summer camps, more and more people were discovering these retreats.

Access is always the key to development. The Pawleys Island causeway, constructed by slaves, was the island's first access road, and additional roads and bridges were added over the years. A railroad was built in 1902, but a 1906 storm destroyed this link. The community on Pawleys Island persisted and grew because of such access, while the island developments not reachable by automobile withered and died (e.g., the Santee Delta islands, North Island, and Debordieu). The connecting highway from Charleston to the Myrtle Beach area completed in 1935 and road improvements made after World War II opened the way for rapid development there. More modern ex-amples include the bridges to Hilton Head Island and Isle of Palms. Access to barrier islands spurs the extension of mainland highway systems; for example, the I-526 bypass around Charleston to Sullivans Island and the planned extension of I-40 from Wilmington, North Carolina, to Myrtle Beach.

Hurricane History: Lessons from the Past

In the late summer of 1686 a Spanish invasion force landed on Edisto Island with the objective of capturing Charles Towne from the English, but a "grievous hurricane" thwarted their effort (appendix D, ref. 9). Although the storm saved the colonists from Spanish swords, it cut a swath of destruction through their settlement, destroying crops, trees, boats, and buildings. The 1686 hurricane was a harbinger of storms to come for the settlers of the South Carolina coast. In the September storms of 1700 and 1713, ships were driven ashore, Charles Towne was flooded, and scores of persons drowned. At least 10 additional storms and hurricanes would affect the South Carolina coast by the end of the eighteenth century, the severest coming in 1752. Also a September storm, the 1752 hurricane generated flood levels 10 feet above the normal high-tide mark and did extensive damage to Charles Towne and its vicinity. The death toll was around 100, including 9 dead on Sullivans Island.

The record of nineteenth-century storms is even more impressive: more than 25 hurri-canes, tropical storms, and northeasters had an impact on the South Carolina coast. The most destructive were the storms of 1804, 1813, 1822, 1854, 1881, 1885, and 1893.

The September 1804 storm made landfall over Beaufort, causing much damage and severe flooding there, but it also washed away Fort Moultrie and 15 to 20 houses at Sullivans Island. Sullivans Island also caught the brunt of the late August storm of 1813. Twelve houses washed away, 15 to 20 people drowned, and the entire island was flooded by 4 to 5 feet of water. Charleston, Edisto Island, and Georgetown were also damaged. Georgetown was flooded again in 1820 by a "September gale," as these storms came to be called. But the late September hurricane of 1822 was the first of the big killers recorded in South Carolina.

The 1822 storm made landfall between Charleston and Georgetown in the Cape Romain area. The stronger right-of-the-eye storm forces drowned several hundred people around Winyah Bay and left between 125 and 300 dead on the North Island plantations. Some of the oldest eyewitness accounts known of storm impact on the barrier islands come to us from this storm. A family on Debordieu watched a North Island house float seaward, lamps still shining from the windows, carrying four victims out into the Atlantic. A family that took refuge from the rising waters in a tree dropped one by one; only two survived (appendix D, refs. 1, 4).

Many of the dead on North Island were slaves, who lived in flimsy houses. As a result of their loss, a number of "storm-proof" towers were built on the island as hurricane refuges for the slaves, perhaps the first storm-conscious construction on the South Carolina coast. Some of these towers were still standing when the big 1893 storm struck. Charleston and Sullivans Island also suffered casualties and destruction in the 1822 storm. Georgetown was lashed again in 1834 and 1837; and the big storm of 1854 did extensive damage from Georgia all the way to Winyah Bay. And so it went through the 1800s. Hundreds of lives were lost in the Savannah area during an August 1881 hurricane. The August 1885 hurricane, concentrated around Charleston, killed 21 people and caused nearly $1.7 million in damage.

C. K. Prevost and E. L. Wilder's book *Pawleys Island . . . a Living Legend* (appendix D, ref. 3) gives a good account of the 1893 storm. The Litchfield Beach (Magnolia Beach) development was hit hard by this storm. One house survived because it stood on high ground; the rest were destroyed, and most of the residents drowned. Survivors estimated wave heights at 40 feet, although it must have been difficult to make a precise estimate when sand and debris-laden wind and water were blinding them and cutting into their flesh. A few lucky ones rode flood debris to the mainland. On nearby Pawleys Island, the ocean rose to meet the creek, and houses on low ground

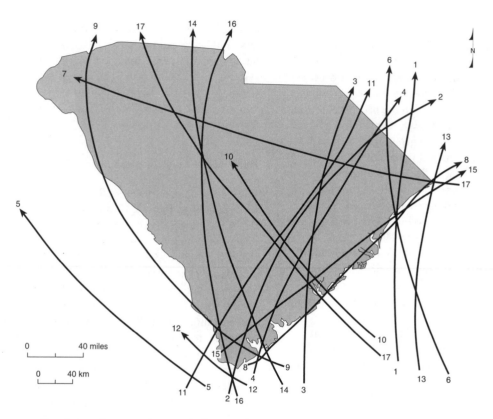

were flooded to the second story. Only two houses were destroyed, however, a success attributed to strong construction and placement on high ground behind high, protective dunes. The ocean borrowed sand from the dunes to flatten the beach. Fortunately, nature had stored enough sand in the dunes for just such a withdrawal, and the dune line held. Farther south, from Charleston to Savannah, the coast

1.7. Hurricane tracks across the South Carolina shoreline, 1883–1989. EH, extreme hurricane; GH, great hurricane; MH, major hurricane. (1) 1883, MH; (2) 1885, EH; (3) 1893, MH; (4) 1894; (5) 1898, EH; (6) 1899, MH; (7) 1906, GH; (8) 1910, GH; (9) 1911, MH; (10) 1916; (11) 1928 GH; (12) 1940, MH; (13) 1954 [Hazel], GH; (14) 1959 [Gracie], MH (15) 1966 [Alma], MH; (16) 1979 [David]; (17) 1989 [Hugo], GH. *Source: South Carolina Hurricanes* (appendix D, ref. 9).

was ravaged. Estimates vary, but at least a thousand lives were lost, and property damage amounted to $10 million. By the time Hugo hit nearly 100 years later, many of the protective dunes were completely gone, more houses filled the island, and the stage was set for greater damage.

Figure 1.7 shows a few late nineteenth-century hurricane tracks as well as tracks of the more destructive hurricanes of the twentieth century (through 1994). Although more than 20 hurricanes affected the state, directly or indirectly, in the twentieth century, only 5 or 6 had a great impact in terms of lives lost and significant property damage. The 1911 hurricane left 17 dead and $1 million in damage in Charleston. The great hurricane of 1928 killed 1,800 in south Florida and came on to strike South Carolina, killing 5 and doing several million dollars' worth of damage. Another killer hurricane struck in August 1940, leaving 34 dead and causing $10 million in losses.

Hurricane Hazel struck the South Carolina coast in October 1954. Hazel made landfall near the state's border with North Carolina, and its greatest impact was on the barrier islands of southern North Carolina. Nevertheless, one death and $27 million in property losses were reported in South Carolina. Myrtle Beach recorded winds of 106 miles per hour, and the tide rose to 16.9 feet above mean low water at Spiveys Swash. Parts of Folly Beach eroded 100 feet; Pawleys Island was breached as the tide rose to nearly 14 feet above mean

1.8. Demolished residence on Folly Island after Hurricane David in 1979. Photograph courtesy of the *Charleston Post-Courier*.

low water and waves inundated parts of the island; 16 cottages were damaged at Litchfield Spit (McKensie Beach); 272 homes were destroyed or heavily damaged at Garden City; the entire front row of cottages at Surfside was lost and second-row houses were badly dam-

aged; and 80 percent of the beachfront property in Myrtle Beach was heavily damaged or destroyed. North Myrtle Beach communities fared no better: 50 of the 89 cottages of the two frontal rows were completely destroyed; 25 houses were lost in Atlantic Beach; 150 houses were destroyed or severely damaged in Crescent Beach; 75 percent of the beachfront houses in Ocean Drive were heavily damaged or destroyed; more than 100 houses were similarly affected at Tilghman Beach; and in Cherry Grove Beach all the front-row houses were destroyed, many on the second row suffered the same fate, and a new inlet was cut by the storm surge (appendix D, ref. 9).

Other hurricanes of the 1950s and 1960s were not as spectacular, passing offshore or striking the shores of neighboring states, but they too cost a few lives and millions in cumulative property losses—a dune here, a boardwalk there—ultimately contributing to the undercutting of someone's beachfront property.

Hurricane David, in August 1979 (fig. 1.8), was a relatively minor offshore hurricane which spent its energy coming ashore, but it caused considerable beach erosion and more than $10 million in damage. Hurricane Hugo, in 1989, was the last great hurricane to strike the South Carolina coast. Hugo is discussed in greater detail in chapter 3 because of its intensity, its impact in property losses, its test of the Beachfront Management Act, and the lessons it offers for living with the coast. As terrible as the storm was, however, there were only a few deaths, indicating that residents received adequate warning. In the weeks and months after the hurricane, according to Dick Shenot, then head of the Charleston office of the National Weather Service, a curious phenomenon was observed. People's memory of the event gradually changed. The entire population seemed to forget its initial horror of Hurricane Hugo, replacing the fear with positive memories of survival and recovery. This is undoubtedly a func-tion of the psychology of survival. Mr. Shenot is concerned that when the next killer hurricane comes along, people will think, "Well, we survived Hugo, so . . ." If he is right, then the psychology of survival may be extremely dangerous. If residents do not evacuate early—or at all—from low-lying barrier islands, particularly those whose escape routes may become bottlenecks, the decades-long downward trend in hurricane death statistics could be reversed.

Northeasters: Winter's Storms

Second to hurricanes in concentrated energy are northeasters (also discussed in chapter 3), winter storms associated with large, intense low-pressure systems that move offshore along the coast and are accompanied by winds and waves out of the northeast. Northeasters typically occur from autumn to spring and usually are of several days' duration. It is not unusual for them to coincide with spring tides. (Spring tides, which occur twice per month, are the highest high tides.) When storm waves are superimposed on these extra-high tides, shoreline erosion is intensified. Perhaps 50 such storms were severe enough to cause some degree of coastal damage along the South Carolina shoreline in the first 80 years of this century. Rarely a year goes by without a hurricane or northeaster eroding some part of the state's shore. (See appendix D for references on specific storms and storms in general.)

Intense northeasters in late October and De-cember 1979 caused considerable erosion along much of the coast, including Myrtle Beach. Possibly these storms were more damaging than usual because the beaches had not had sufficient time to rebuild from the effects of Hurricane David. A northeaster that struck on December 1, 1986, packed enough force to erode through the washout at Folly Beach. The washout was rebuilt and the road repaired, but it was destroyed again exactly one month later, on January 1, 1987. Other memorable northeasters include the Halloween storm of 1991 and the 1993 "Storm of the Century."

One thing is certain from a review of storm history: storms do not occur in a regular pattern or with regular spacing. One big storm can follow on the heels of another, or several years can go by without a big event. The only certainty is that several storms will affect a house over its lifetime. In order to prepare for such storms and attempt to reduce their impact in terms of property losses and potential loss of life, we must first understand how barrier islands work and the potential for conflict between nature and development.

Lessons for Coastal Management

Poststorm observations of the impacts of Hurricanes Gilbert (1988), Hugo (1989), Bob (1991), Andrew (1992), Emily (1993), and several winter storms have helped to define key principles for living with the South Carolina coast (appendix D, refs. 92, 93). We conclude this chapter and set the stage for those that follow by listing 10 lessons we have learned from these storms.

1. *Wide beaches protect property.* The more beach available to absorb and dissipate storm wave energy, the greater the protection for developed areas.

2. *Dunes protect property.* Sand dunes are often referred to as the "barriers" in barrier islands and "nature's shock absorbers." The mass of dune sand may absorb and dissipate storm-wave energy, thus protecting buildings located behind dunes.

3. *Vegetation protects property.* Overwash penetration and storm damage are noticeably greater where vegetation, especially maritime forest, has been removed for development. A stabilizing cover of dune grass, marsh grass, maritime forest, and even lawn and landscaping is important to each respective environment.

4. *Shore-perpendicular roads act as overwash and storm-surge ebb conduits.* Elevating and curving roads so that they approach the beach at an oblique angle reduces the extent and amount of overwash and damage from ebb flow.

5. *Notches in dunes create overwash passes.* Notches cut in dunes for beach access, views, or construction sites are naturally exploited by waves and storm surge, and by storm-surge ebb flows.

Notching can be avoided by constructing walkovers, elevating structures, and generally protecting frontal dunes.

6. *Overwash and storm-surge ebb are intensified when funneled by structures.* The force of the storm-surge waters that overwash an island and then return to the sea is greatly intensified when the waters are constricted between structures. The greater the density of buildings, the worse this problem becomes.

7. *Seawalls can protect buildings, but they also can cause narrowing of the beach in both a recreational and a protective sense.* Large seawalls are effective for protecting shorefront buildings from wave attack. But seawalls, as a rule, cause degradation and even eventual loss of beaches.

8. *Setbacks protect.* Choosing a building site well back from the sea is the easiest and least costly method of protecting development.

9. *Elevation protects.* Elevation, whether achieved by natural land elevation, infilling on a construction site, or building on pilings, may be the single most important site-specific factor in reducing property damage.

10. *Proper development offers a degree of self-protection.* Development where building codes are enforced and barrier island processes are allowed to operate is less susceptible to property damage.

These principles must be applied islandwide and all together; they are inadequate if applied singly or in one subenvironment. Although recommended actions may be site-specific, these principles must be compatible and applied over the entire island system.

If you are planning to live on or visit one of South Carolina's barrier islands, you should understand the natural processes at work there. The well-being of the island and your own safety are at stake. Structures built in the coastal zone must be able to withstand these natural processes. Even the Myrtle Beach area, which is now mainland coast, originated as part of an ancient barrier island system, and community planning and site-specific analysis there can be viewed according to the principles of barrier island dynamics.

Barrier Islands

Where Did They Come From?

The forces acting on the islands today are similar to those that created them. Thus, knowing how the islands were formed will help you to appreciate the modern forces.

South Carolina has barrier islands because the rising sea level interacts with a coastal plain indented by river valleys (fig. 2.1.). Approximately 15,000 years ago, when the sea level was as much as 150 feet lower than it is today, the South Carolina shoreline was approximately 50 miles offshore, on what is now the continental shelf (fig. 2.2). Vast glaciers covered the high latitudes of the world, tying up a great deal of water. When the climate became warmer, the ice started melting and the sea began to rise. The rising water flooded the river valleys, forming bodies of water called

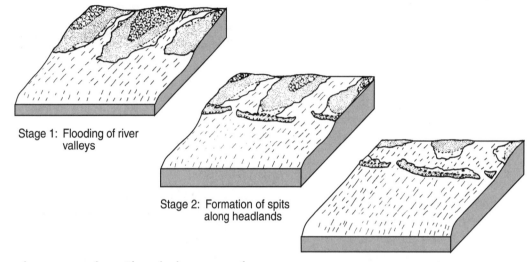

Stage 1: Flooding of river valleys

Stage 2: Formation of spits along headlands

Stage 3: Separation of barrier from mainland

embayments, or bays. If you look at a map of today's shoreline, you can see many such inundated valleys, especially along the Atlantic coast of the United States. Chesapeake Bay and Delaware Bay are two prominent examples. Port Royal Sound, St. Helena Sound, Charleston Harbor, and Winyah Bay are examples found in South Carolina.

If this inundation were all that had occurred, the shoreline today would be jagged. Nature, however, tends to straighten out jagged shorelines. Wave action cut back the headlands—the areas of land that extended seaward between flooded valleys—and built spits, thin fingers of sand extending from the headlands across the bay mouths. As the sea level contin-

2.1. Evolution of barrier islands. Bays develop first, then spits form from sand delivered by erosion of headlands between the bays. Finally, the rising sea level floods the land behind the spits and isolates them to form barrier islands.

ued to rise, the low-lying land behind such spits became flooded. Eventually, the dune-beach complexes were detached from the mainland, and the barrier islands were born.

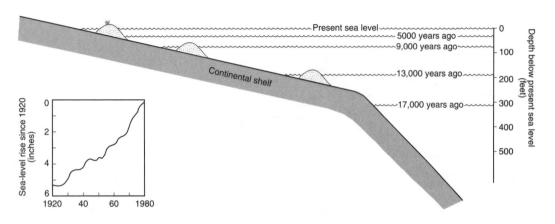

2.2. The rise in sea level during the past 17,000 years. Barrier island migration is driven by the rising sea level.

Islands on the Move

You might ask why, if the sea level continued to rise, the newly formed islands were not themselves covered by the sea. The answer is this: when the level of the sea rises, barrier islands do not stay in one place. They move—scientists say "migrate"—toward the mainland. The more rapid the rise in sea level, the faster they move. Needless to say, in order for the islands to have remained islands, the mainland shore must have moved, too. Otherwise the islands would have run aground (somewhat like what happened in the Myrtle Beach area). As the sea level continued to rise and the sea advanced over the land, the shoreline on the mainland retreated.

The sea level rise was quite rapid until about 5,000 years ago, when it slowed down considerably (fig. 2.2). Hence, up until 5,000 years ago South Carolina's barrier islands were moving landward at an impressive clip. Rapidly moving islands tend to be low, very narrow strips of sand and are referred to as *transgressive* islands. Cape Island (fig. 2.3) is a small-scale example of a transgressive island, whereas portions of North Carolina's Outer Banks are examples on a large scale.

When the rise in sea level slowed, many islands stopped migrating altogether. They also began to widen then, because they remained in one position long enough for sand from various sources to accumulate. Beach-dune complexes built successive ridges in the seaward direction, giving islands such as Hilton Head, Kiawah, and Seabrook their characteristic ridge-and-swale topography. This relative stability, however, has recently come to an end.

The Accelerating Rise in Sea Level

In the 1930s the rise in sea level suddenly accelerated. The sea level is now rising at a rate of perhaps slightly more than 1 foot per century. Keep in mind that this refers to a vertical rise. The horizontal change—and the distance islands migrate as a consequence—is much greater, between 100 and 1,500 feet per century. How much a specific island moves depends on the slope of its migration surface: the gentler the slope, the more the island will migrate (fig. 2.4).

The safest assumption you can make about the future of the sea level rise is that it will continue and accelerate. The National Academy of Sciences has warned that the earth's surface is becoming warmer. The burning of fossil fuels

2.3. Cape Island, a narrow transgressive island that is easily overwashed during storms. Note the lobe-shaped marsh, once a flood-tidal delta.

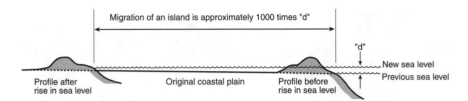

Migration of an island is approximately 1000 times "d"

"d"

New sea level
Previous sea level

Profile after
rise in sea level

Original coastal plain

Profile before
rise in sea level

2.4. Ratio of horizontal island migration to vertical sea level rise.

has resulted in the excessive production of carbon dioxide, which causes the atmosphere to retain heat. This warming is expected to increase the rate at which the polar ice caps melt, which in turn will accelerate the sea level rise.

Barrier Island Migration

Do you want to prove to yourself that barrier islands migrate? If you happen to be standing on one now, walk to the ocean-side beach and look at the seashells. Chances are that on most South Carolina beaches you will find oyster, clam, or snail shells that once lived in the waters off the back side of the barrier island, the side facing the mainland. How did shells from the back side get to the front side? The answer is that the island migrated over the back-side area, and waves attacking and breaking up the old back-side sands and muds threw the shells up onto the present-day beach. Of course, this assumes that you are looking at a natural beach and that the sand has not been pumped in from behind the island as artificial nourish-

ment (see chapter 4).

Many of the shells found on Atlantic beaches are thousands of years old. In addition, salt-marsh peats that formed on the back sides of the islands at earlier times are occasionally exposed on ocean-side beaches after storms (fig. 2.5); this happens, for instance, on Hilton Head Island, Botany Bay Beach, Folly Beach, and beaches of the Cape Romain area. A patch of mud that appeared on the beach at Whale Beach, New Jersey, after a storm contained (much to the surprise of some beach strollers) cow hooves and fragments of colonial pottery. The mud had once been a salt marsh on the back side of the island where a colonist had dumped a wagonload of garbage. Since colonial times this particular section of the island had migrated over its entire width! The same thing occurs on the northern part of Folly Island, formerly Little Folly Island, where shoes, broken wooden tools, animal bones, and other debris have appeared on the front beach.

If you haven't guessed already, *island migration* is the term geologists use for what beach homeowners call *beach erosion.* In order for an island to migrate, the side facing the ocean

(the front side) must move landward by erosion, and the side facing the mainland (the back side) must do likewise by growth. Also, as it moves, the island must somehow maintain its elevation and bulk (fig. 2.6).

The Front Side Moves Back by Erosion

The beach on the front side moves back because the sea level is rising. The rise in sea level is the main cause of beach erosion worldwide, although human intervention can make erosion worse—sometimes much worse—by interfering with the sand supply. Examples in South Carolina include erosion caused by the engineering of the Santee River and the Charleston Harbor jetties (see chapter 6).

Shore-hardening structures inhibit the island thinning process, either by reducing front-side erosion (e.g., seawalls) or by causing beach widening (e.g., accretion on the updrift side of jetties, as on Sullivans Island and Isle of Palms). As a result, the island falls out of equilibrium with the migratory processes.

The Back Side Moves Back by Growth

There are several ways for an island to widen. It may grow on the seaward side by the accretion of sand into dunes, a process that happens either as a result of falling sea level or because of the influx of a large quantity of sediment at a time when sea level rise is not great. (Need-

2.5. Left: Salt marsh deposit exposed on a beach. Because salt marshes form behind islands, this outcrop is certain evidence of island migration.

2.6. Right: Island migration along Edingsville Beach. The marsh behind the island is filling in. Old salt marsh mud is exposed on the beach as the island retreats into and over the marsh. Note that a beach is present even though the shoreline is eroding. The back side of the island is widened as the result of overwash fans (white areas).

less to say, this situation is not a common luxury in South Carolina today.)

When sea levels are rising, it is much more common for barrier islands to grow on the back, or landward, side. One of the processes by which barrier islands grow landward is overwash. During storms, waves frequently penetrate through shorefront dunes and deposit sand bodies called overwash fans. If beach erosion is great or the frontal dune line is weak, these overwash fans may coalesce to form a washover terrace. On narrow islands, overwash fans or terraces may reach all the way to the marsh on the landward side of the island. Conversely, if the frontal dune is continuous and massive, it may prevent washover sand from reaching the back side of the island. Thus, the process of overwash adds both to the existing high ground and to the marsh at the back of the island. In a natural system, an island retains its equilibrium, and thus its general size and shape, while migrating landward. Erosion of the beach side is compensated for by growth on the landward side.

Overwash is the method of back-side growth used by islands in a hurry—that is, those that are migrating rapidly landward. Cape Island in South Carolina and some of Louisiana's islands are examples. From 15,000 to about 5,000 years ago, when the sea level was rising rapidly, most U.S. barrier islands were probably of the overwash type.

Another way islands grow landward is through the incorporation of flood-tidal deltas. Deltas—fan-shaped lobes of sediment—invariably form both seaward and landward of inlets. Based on the tidal conditions and other

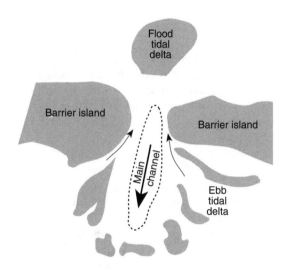

2.7. A typical configuration of sandbars forming the flood-tidal and ebb-tidal deltas on a barrier island. If the inlet closes, the flood-tidal delta becomes part of the now wider island, and the sand of the ebb-tidal delta will be dispersed by the longshore currents.

factors, an inlet may be dominantly ebb-tidal or flood-tidal (fig. 2.7). Inlets in South Carolina tend to be ebb-tidal, meaning that they produce deltas predominantly seaward of the inlets. Nevertheless, flood-tidal deltas, although smaller, are part of the inlet systems associated with our barrier islands. As inlets migrate laterally between the islands, the location of these flood-tidal deltas also moves. Barrier islands moving in response to rising sea levels eventually encounter these abandoned flood-tidal deltas and incorporate them into their landward sides.

The Island Maintains Its Elevation during Migration

The remaining problem for a migrating island is how to retain its bulk or elevation as it moves toward the mainland. This problem is solved by two processes: dune formation and overwash fan deposition.

Dunes are formed by the wind. If waves bring a sufficiently large supply of sand to the beach from the continental shelf, a high-elevation island can be formed. The lack of a sand supply or a dominant wind direction blowing up and down the beach rather than across it can prevent dune formation on islands of low elevation.

Gaps between dunes allow overwash to penetrate into the island's interior, and sometimes to its back side. The sand accumulates and builds elevation. If the dunes are destroyed, overwash gain can become the dominant process. In contrast, if a continuous artificial dune line is constructed, overwash is blocked and the beach narrows. And the island's interior, behind the dune or other artificial barrier, is deprived of its nourishment.

Narrow Islands Migrate Most Rapidly

A note of caution about the processes of migration described above: do not assume that the back of any barrier island is growing. Many barrier islands today, especially those with a sound rather than a marsh behind them, are eroding on both the front and back sides. This is because effective and rapid migration can be carried out only by narrow islands, like Cape Island (fig. 2.3), and many islands are presently too wide to respond to the rise in sea level in the way just discussed; these islands are thinning down in preparation for more rapid migration. Meandering tidal creeks also may erode the back side of an island.

The Size and Shape of South Carolina's Barrier Islands

Several studies of South Carolina islands are listed under "Barrier Island and Beaches" in appendix D. The size and shape of the barrier islands in the Carolinas are functions of the tidal range—the difference between water levels at high and low tides (see table 2.1). South Carolina lies in the transition zone between the microtidal range (0–6.5 feet) of the North Carolina coast and the mesotidal range (6.5–13 feet) of the Georgia Embayment. North Carolina's Outer Banks are typical microtidal barrier islands: they are relatively long, low in elevation, frequently overwashed, and cut by few inlets. The inlets feature flood-tidal deltas that are more prominent than the ebb-tidal deltas.

Typically, South Carolina islands show the effects of the higher tidal range. Most are less

Table 2.1   Tidal Range and Barrier Islands

Tidal Range	Microtidal (0–6 feet)	Mesotidal (6–13 feet)	Macrotidal (>13 feet)
Island size and shape	Long (18–60 miles); elongated "hot dog"	Short (<2–12 miles); "drumstick" shape	Barrier islands absent; embayments
Overwash	Abundant; fans common	Less common; beach ridges block overwash	—
Inlets	Infrequent, sometimes called "swashes"	Numerous and/or very large	—
Tidal deltas	Large flood-tidal deltas; ebb-tidal deltas small or absent	Large ebb-tidal deltas influence wave patterns; flood-tidal deltas of moderate size or absent	Extensive tidal flats and salt marshes
Back (behind) barrier	Narrow lagoons or marshes	Large estuaries, very large, well-developed marshes drained by tidal creeks	Tidal mudflats
Examples	Pawleys	Folly, Kiawah	None in S.C., Bay of Fundy, Canada

Sources: *Historical Inlet Atlas for South Carolina* (appendix D, ref. 59); and *Terrigenous Clastic Depositional Environments* (appendix D, ref 63).

than 12 miles long and usually have a well-developed beach ridge (a row of sand dunes parallel to the beach) sufficient to block overwash. These islands are backed by broad, well-developed marshes. South Carolina's barrier islands are cut by numerous inlets whose ebb-tidal deltas are larger than their flood-tidal deltas. Two main barrier island types are referred to as the "hot dog" model and the "drumstick" model. Maps and aerial views of Bulls Island, Kiawah Island, and Folly Island show their characteristic drumstick shape.

Figure 2.7 presents a model of an ebb-tidal delta of the type commonly associated with South Carolina inlets. These deltas, which store large volumes of sand that is occasionally added to adjacent islands, protrude seaward and interfere with waves that are approaching the shore at an oblique angle. Such waves normally produce a current, called a longshore current, that moves north to south along the shore (fig. 2.8). The deltas, however, refract these waves and cause the longshore current south of the delta to run in the opposite direction, south to north.

The reversed current south of the deltas plus the occasional pileup of sand from the deltas onto the adjacent beaches account for the "fat" end of the drumstick and the tendency for the island to stick out farther seaward on the south side of the inlet. So, typical South Carolina islands are fat to the north and thinner to the south. Natural or artificial removal of a delta can change the shape of the island adjacent to it.

The narrow, elongated end of the drumstick is the result of sand-spit growth into inlets that are migrating southward. Inlet migration changes with the rate and direction of material carried by the longshore current, which varies with storm history, depth of the inlet, and sand supply. These variables will themselves change in response to both natural and man-made alterations of an island (appendix D, ref. 118).

South Carolina islands less than 4 miles long tend to be under the influence of their adjacent

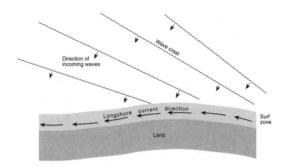

2.8. Longshore currents are formed by waves approaching the shoreline at an angle. The longshore current transports sand parallel to the shore. The wave approach and backwash also carry beach sand parallel to the shore (beach drift).

mobile inlets over their entire length. Such islands are too dynamic to be suitable for development (e.g., Waites Island and Dewees Island). The central portion of the larger islands is usually stable enough for development, but the ends of such islands may suffer from inlet migration or from the potential for new inlet formation on the narrow, spit end. Much of the severe erosion and associated property loss now experienced in South Carolina occurs in the vicinity of tidal inlets (e.g., Seabrook and the washout on Folly Beach).

A Classification of the South Carolina Coast

South Carolina's coast can be classified into four basic landform types (fig. 2.9; appendix D, ref. 63, 118). This classification can be used in conjunction with the individual island and specific site analyses given in chapter 6 to evaluate the relative risk to development from coastal hazards at locations along the South Carolina coast. The four landform types are *arcuate strand, delta, beach-ridge barriers,* and *transgressive barriers.* We use the less technical terms *mainland coast, delta, barrier with sand dune ridges,* and *thin, retreating barriers,* respectively, here.

Mainland Coast (Arcuate Strand)

The shoreline from the North Carolina line to Winyah Bay is characterized by a wide, curved

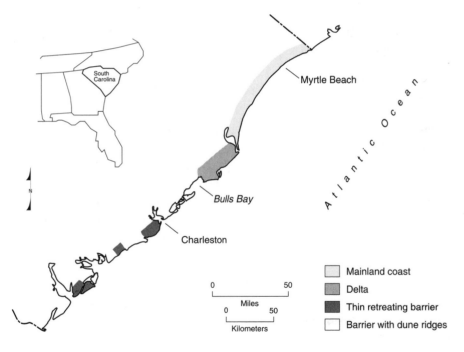

2.9. Classification of the South Carolina coast. *Source: Terrigenous Clastic Depositional Environments* (appendix D, ref. 63).

beach cut by a few small inlets. The shore zone is part of an ancient Ice Age barrier island complex stranded on a plateau known as the Princess Ann Terrace. This part of the coast is generally regarded as the most stable; however, erosion is a reality even here, and it is compounded in places by dense development.

Erosion may be accelerating in this area partly as a result of the type of development now occurring. Beach sand is derived from the older Ice Age deposits lying directly behind the shoreline. As these sand deposits have been covered by buildings, streets, and parking lots, or separated from the beach by walls and revetments (see chapter 4 on shoreline engineering structures), the sand supply available to the beach during storms has been reduced. Walls and revetments also reflect wave energy, hastening the offshore transport of existing beach sand. Thus, this area, which might have been the most appropriate stretch of the South Carolina coast for development, has not been developed in keeping with the natural character of the coast.

Delta

The Santee River Delta (Cape Romain) is the largest delta complex on the east coast of the United States. Extensive salt marsh covers the lower delta plain. The sediment making up the delta was derived from the Santee River, but construction of dams upstream and the diversion of much of the Santee's flow into the Coo-

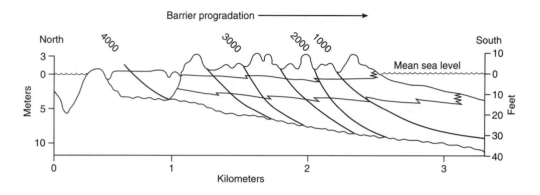

2.10. Generalized cross section through the central portion of Kiawah Island. The numbers are time lines constructed on the basis of radiocarbon-dated samples taken from cores. Kiawah Island has experienced progradation over the last 4,000 years.

per River in 1942 cut off the sediment supply, causing widespread erosion. The river flow was redirected in the 1980s, but only the water is channelized back into the Santee; the sediments are left behind, far from the coast. Therefore, massive erosion will continue here. This coast is totally unsuited to development.

Barrier with Sand Dune Ridges (Beach-Ridge Barriers)

This coastal type generally occurs on the drumstick-shaped islands (e.g., Capers Island, Bulls Island, Isle of Palms, Folly Island, Sullivans Island, Kiawah Island, and the south-

ern half of Hilton Head Island). Typically, well-developed, vegetated beach ridges (rows of sand dunes originally formed adjacent to the beach) make up the stable interior parts of these islands. The ridges formed when the sea level rise slowed about 5,000 years ago. The slower rise in sea level meant that more sediment was carried onshore than was eroded away. The islands thus built seaward with a series of beach ridges welding themselves onto the primary barrier (fig. 2.10; appendix D, ref. 119). The central portions of drumstick islands more than 4 miles long can usually be regarded as marginally suitable for development provided the site is well behind the frontal dune barrier. Areas in the proximity of inlets (within about 0.5 mile) and islands less than 4 miles long (e.g., Dewees Island and Harbor Island) should be avoided, as noted earlier.

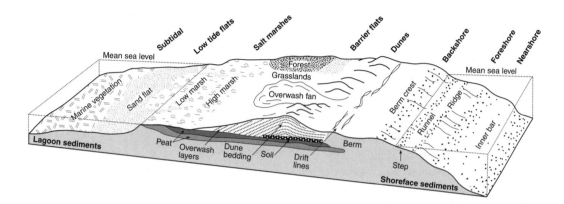

The image shows a three-dimensional block diagram of barrier island environments with labels across the top: Subtidal, Low tide flats, Salt marshes, Barrier flats, Dunes, Backshore, Foreshore, Nearshore. Other labels include: Mean sea level, Forest, Grasslands, Marine vegetation, Sand flat, Low marsh, High marsh, Overwash fan, Berm crest, Ridge, Runnel, Inner bar, Mean sea level, Lagoon sediments, Peat, Overwash layers, Dune bedding, Soil, Drift lines, Berm, Step, Shoreface sediments.

2.11. Barrier island environments. *Source:* "Barrier Beaches of the East Coast" (appendix D, ref. 71).

Thin, Retreating Barriers

Several of South Carolina's barrier islands are not drumstick types (e.g., Morris Island, Edingsville Beach, and Bay Point), but instead are thin strips of sand with straight shores that are rapidly migrating landward. As the island migrates, beach erosion exposes peats and marsh muds that accumulated on the former back side of the island. The sands are shelly, typically including oyster-shell debris. Dunes and beach ridges are absent. Like the Cape Romain beaches, these islands are totally unsafe for development. It is of historic interest that both Morris Island and Edingsville Beach once had beach ridges! Islands can change not only in position but in type.

**Barrier Island Environments:
An Integrated System**

It is important to understand that the environments found on each barrier island are interrelated (fig. 2.11). Each environment is part of an overall integrated system and to some degree depends on or affects other environments within the system. Specific environments are discussed in chapter 5.

Perhaps the best example of one environment affecting others in the system is provided by the role of the ocean-side beach. The beach is important both because it alters its shape during storms in a way that minimizes fundamental damage to the island by waves, and because it is the major source of sand for the entire island. Examples of the ways humans have interfered in the integrated system may best illustrate these functions.

As we discussed earlier, overwash is one of the processes by which barrier islands grow and migrate landward. During storms, waves frequently penetrate shorefront dunes and deposit overwash fans. On narrow islands, washover fans or terraces may reach all the way to the marsh on the landward side of the island. Conversely, if the frontal dune is continuous and massive, it may prevent washover sand from reaching the back side of the island.

The integrity of a primary dune is sometimes artificially "improved" to reduce beach erosion and island migration. However, this seemingly benign shore engineering causes erosion on the landward side of the island because it prevents overwash. This can be a problem if the island is very narrow (and thus is actively migrating) and is backed by open waters of a lagoon or estuary. Barrier island processes are intimately linked. We cannot assume that our interference with one process, no matter how good our intentions, will not have adverse consequences elsewhere on the island.

The maritime forest also illustrates the integration of island environments. The larger trees are salt-tolerant and form a canopy over the less tolerant undergrowth. The undergrowth, in turn, stabilizes the larger trees by holding down the soil. If trees are thinned or removed, salt spray can attack and eliminate the undergrowth, which in turn allows sediment to be eroded by wind or other processes, thereby destroying the larger trees. Figure 2.12 shows good site selection that takes advantage of the forest canopy and retains as much of the protective forest buffer as possible.

Much has been said about the damage to islands caused by dune buggies and other off-road vehicles. Dune buggies can prevent dunes from stabilizing (becoming stationary), and destabilization (moving sand) may result in destroyed dunes and loss of vegetation as well as sand dune migration into maritime forests or developments. South Carolina no longer allows driving on its beaches. And, of course, destruction of oceanfront dunes is prohibited.

The most common cause of excessive sand movement on barrier islands is construction. The problem is particularly acute during the early stages of construction, and in many in-

stances has halted construction altogether. Poorly designed roads are another common mistake. On many American barrier islands you can drive along roads that parallel the beach and observe that at the end of each feeder road to the beach there is a giant notch through the last row or two of dunes. Such notches provide a path for storm-wave overwash and the strong currents of return ebb flow during hurricanes and northeasters. Beach access roads should go over, not through, dunes, and should curve away from the beach rather than cutting straight across the island (appendix D, ref. 98).

Just as the environments on a single island depend on one another, so do the environments on adjacent islands. The beaches on our islands are flowing rivers of sand. Frequently islands depend on neighboring islands for their

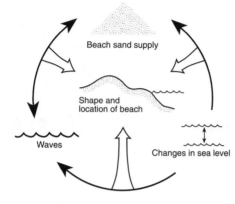

2.13. The dynamic equilibrium of beach shape and position.

sand supply, sometimes via ebb-tidal deltas. When this sediment supply is cut off by inlet dredging or jetties, the adjacent island's frontal erosion rate increases.

2.12. Houses constructed in maritime thicket and forest can conserve protective cover. The inner island site, elevation, and natural protection reduce the property's vulnerability to hazards.

Beaches: The Dynamic Equilibrium

The beach—the zone of active sand movement—is one of the earth's most dynamic environments. It is always changing and always migrating, and we now know that it does so in accordance with natural laws, which control a beautiful, logical environment. The beach builds up when the weather is good, and strategically (but only temporarily) retreats when confronted by big storm waves. This system depends on four factors: waves, sea level rise, beach sand supply, and the shape and position of the beach (fig. 2.13). The relationship among these factors is a natural balance referred to as a "dynamic equilibrium": when one factor changes, the others adjust accordingly to maintain the balance. When humans enter the system incorrectly—as we often do—the dynamic equilibrium continues to function, but in a way that is harmful to the system.

The answers to several frequently asked questions about beaches clarify the nature of this dynamic equilibrium. Keep in mind that the beach extends from the toe of the dune offshore to a water depth of 30 to 40 feet (often a mile or several miles out). The entire beach is the zone of sand movement during storms. The part on which we walk is only the upper beach.

How Does the Beach Respond to a Storm?

Old-timers and survivors of storms on barrier islands frequently comment on how flat and broad the beach is after a storm. The flat beach can be explained in terms of the dynamic equilibrium: as wave energy increases during a storm, the dunes at the back of the beach are eroded (scarped) and sand is moved across the beach, changing its shape (profile). The reason for this storm response is logical. When the beach is flatter, the storm waves must expend their energy over a broader and more level surface. On a steeper surface, storm-wave energy would be expended on a smaller area, causing greater damage. Figure 2.14 illustrates the way

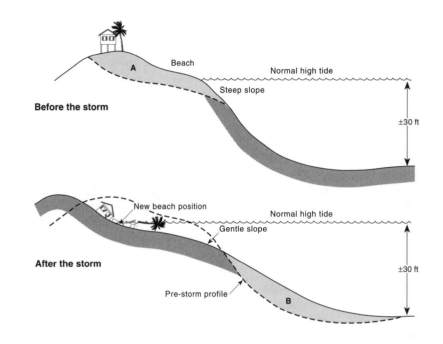

2.14. Beach flattening in response to a storm. Shaded area of B is approximately equal to A.

the beach flattens. To summarize what happens, the waves take sand from the upper beach, or the first dune, and transport it to the lower beach. If a beach house happens to be located on the first dune, it may disappear along with the dune sands.

A great deal of island sand may be moved offshore during a storm, but much of it will come back, gradually pushed shoreward by fair weather (typically summer) waves. As the sand returns to the beach, the wind takes over and slowly rebuilds the dunes, storing sand to respond to the call of nature's next storm. In order for the sand to come back, of course, there must be no structural obstructions—such as a seawall—between the first dune and the beach. The return of the beach may take months or even years, and if houses are built in the meantime where the dunes should be, this natural protection will not be restored. The construction setback line should be behind the dune line—either the healthy existing dunes or where the line will reform if the dunes have been destroyed by a storm.

Besides simply flattening, a storm beach also develops one or more offshore bars. The bars trip the large waves long before they reach the beach, thus dissipating their energy and minimizing damage. A sandbar produced by storms is easily visible during calm weather as a line of surf a few yards to tens of yards off the beach. Geologists refer to the bar as a "ridge" and the intervening trough as a "runnel" (fig. 2.15).

How Does the Beach Widen?

Beaches grow seaward in several ways, but principally by bringing in new sand via the so-called longshore (surf zone) currents and by bringing in new sand from offshore by forming a ridge-and-runnel system. Actually, these two ways of beach widening are not mutually exclusive.

Longshore currents are familiar to anyone who has swum in the ocean; they are the reason you sometimes end up way down the beach, far away from your towel. We commonly refer to this phenomenon as "undertow." Such currents result when waves approach the beach at an angle and a portion of the energy of the breaking wave is directed along the beach. When combined with breaking waves, the longshore current is capable of carrying large amounts of very coarse material for miles along the beach. The sand transported along the shore may be deposited at the end of the island, causing the spit growth so common on the south ends of South Carolina's islands.

Ridges and runnels (fig. 2.15) formed during small summer storms virtually march onto the shore and are "welded" to the beach. The next time you are at the beach, observe the offshore ridge for a period of a few days and verify this for yourself. You may find that each day you have to swim out a slightly shorter distance in order to stand on the sandbar.

At low tide during the summer, the beach frequently has a trough filled or partly filled with water. This trough is formed by the ridge that is in the final stages of welding onto the beach. Several ridges combine to make the berm, or beach terrace, on which sunbathers stroll.

Where Does Beach Sand Come From?

Along the west coast of Florida and along most of the Atlantic portion of the American barrier coast—which runs approximately 10,000 miles, from the south shore of Long Island, New York, down and around to where the Texas coast meets Mexico—the sand comes from the adjacent continental shelf. It is pushed up to the beach by fair-weather waves. Additional sand, sometimes in very large quantities, is carried laterally by longshore currents that move in the surf zone, parallel to the beach. Rivers do not generally contribute sand directly to the Atlantic coast's barrier beaches. River sands are deposited far inland at the heads of the estuaries or behind dams.

Beach dwellers should know the source of the sand for their beach. If, for example, there is a lot of transported longshore sand in front of your favorite beach, the beach may well disappear if someone builds a groin "upstream." Is the source of your beach's sand finite or tied to replenishment from an adjacent island? Community actions taken on an adjacent island or inlet potentially could affect your beach, just as your actions may affect your coastal neighbors. Actions taken years ago are still affecting downdrift beaches. The fact that 25 percent of South Carolina's developed shoreline has engineering structures means that human-induced erosion problems are still causing beach loss.

Where Do Seashells Come From?

Surprisingly, most of the shells on many barrier island beaches can be called fossils. Radiocarbon dating has shown that many South Carolina beach shells are between 7,000 and 9,000 years old. Even some of the shiny lettered olive and pretty whelk shells are very old.

If you use a guidebook to identify the specimens you pick up on the beach, you will find that shells originally from the back side of the

2.15. Left: This ridge-and-runnel system on North Myrtle Beach has three distinct ridges separated by two runnels.

2.16. Right: Healthy marsh shoreline on the back side of Folly Island. Note the revetted section of the shore and the healthy marsh growing in front of the rocks.

island or sound are very common on the ocean-side beach. This is because the islands have migrated landward over the shells of animals that once lived in the back-island environments. After a few hundred or a thousand years, these nonocean shells were washed out onto the ocean-side beach. Still, as any beach buff knows, not all seashells are fossils by any

means. Coquina clams live in the upper beach and hasten to rebury themselves when exposed by sand castle builders. Their skeletons also contribute to beach sediment. When we nourish our beaches with sands from a variety of sources (onshore, offshore, and lagoon), we end up with a confusing mixture of shells. And when communities of shelled animals are killed (e.g., by pollution or environmental change), another small contributor to beach sediment supply is lost.

Why Do Beaches Erode?

As we have already pointed out, beach erosion is the common name for the larger process

called island or shoreline migration. Landward shoreline migration is a change induced in the equilibrium profile (fig. 2.13). When sea level rises, wave height increases and the beach sand supply is reduced. The principal cause of long-term erosion is the rising sea level—which currently averages about 1 foot per century along U.S. shores, though some evidence indicates higher rates locally. Sea level rise can differ in different coastal areas because the land also may be slowly sinking or rising relative to the sea level. We in the Carolinas can be thankful that we don't have the more rapid 3-feet-per-century rise of the New England coast.

Working in conjunction with the rise in sea level are the many forces of nature acting at the shore, including waves, tides, and winds. These forces are aided and abetted by the effects of structures that we create: jetties and groins, houses blocking overwash, dams on rivers, and perhaps even the roads that provide us with access to the beach.

If Most Ocean Shorelines Are Eroding, What Is the Long-Range Future of Beach Development?

The long-range future of beach development will depend on how individual communities respond to their migrating shoreline. Those communities that choose to protect their front-side houses at all costs need only look to portions of the New Jersey shore to see the end result. The life span of houses can unquestionably be extended by "stabilizing" a beach (slowing the erosion). The ultimate cost of slowing erosion, however, is loss of the beach. The length of time required for destruction of the beach is highly variable and depends on the shoreline or island dynamics. Usually an extensive seawall on a barrier island will do the trick in 10 to 30 years. Often a single storm permanently removes the beach in front of a seawall.

Building artificial beaches—an alternative to seawalls—also gives diminishing returns in terms of property protection. Each successive generation of the artificial beach may be more expensive and shorter-lived, and the reserves of renourishment sand are finite.

If, when the time comes, a community grits its teeth and either moves the front row of buildings or lets them fall into the ocean, the beaches can be saved in the long run. Unfortunately, the primary factor involved in shoreline decisions in the United States, decisions every beach community must sooner or later make, is money. Saving the beach is of secondary importance to those who own property on the beachfront. Lawyers and consultants working for this loud and vocal minority will use every tool available, including political pressure, to "protect" (read "inflate") property values. When all these measures are added to the balance sheet, shoreline engineering may be cost-justified. Poor communities let the island roll on. Rich ones attempt to stop it. The future of shoreline development in the United States appears to be increasing expenditures leading to the increasing loss of beach.

Are the Shorelines on the Back Sides of Our Islands Eroding?

Most of the South Carolina's islands have healthy salt marsh growing on the landward side, protecting the back side from erosion (fig. 2.16). If a sand bluff, surf-zone stumps, or marsh-mud scarp appears on the back side, however, beware—erosion is occurring. Rates of erosion have not been determined for most back-island shorelines.

What Can I Do about My Eroding Beach?

This question has no simple answer, but it is briefly addressed in chapter 4. If you are talking about an open-ocean shoreline, there is nothing you can do unless (1) you are wealthy or (2) the U.S. Army Corps of Engineers steps in. Your best response, from an environmental standpoint as well as from the long-term economic point of view, is to move your threatened house elsewhere. The bottom line is that the methods employed in trying to stop erosion on an open-ocean shoreline ultimately increase the erosion rate. For example, having a friendly bulldozer operator push sand up from the lower beach will simply steepen your beach's profile and cause it to erode more rapidly during the next storm. This has been tested many times in South Carolina and is no longer permitted. Pumping in new sand (replenishment) costs a great deal of money, and in most cases the artificial beach will disappear much more rapidly than its natural predecessor.

There are many ways to stop erosion in the short run if sufficient money is available, but in the long run, erosion cannot be halted. Coastal residents face one of nature's greatest and most powerful agents: the sea and her storms!

Hurricanes, billed as "the greatest storms on Earth," are certainties on the South Carolina coast. Hurricane Hugo (fig. 3.1) was one of the most cursed and costliest, but it was also one of the most interesting and best studied (see appendix D), and we know a great deal more about how the coast responds to hurricanes because of it. The coast of the Carolinas is a natural laboratory for the living experiment of increasing development in areas at high risk to damage from hurricanes and winter storms (northeasters).

A Word about Hurricanes

Sea level rise may be the ultimate villain in the shoreline saga, but hurricanes are the most memorable actors. Coastal processes of wind, waves, storm surge (the increase in water level during a storm), and overwash are greatest during hurricanes, but the majority of today's coastal residents and property owners have not experienced such storms. The relatively hurricane-free period from the 1960s through the 1980s (before Hugo) contributed both to apathetic disregard of the hurricane menace and to unchecked development in high-hazard zones. Time was not on the side of such development, and the clock is again counting down to the fate of the post-Hugo redevelopment, which is bigger and at even greater risk.

Each year on June 1 the official hurricane season begins. For the next five to six months, conditions favorable to hurricane formation can develop over the tropical and subtropical ocean waters of the Western Hemisphere. The hurricanes that strike the eastern United States early in the season tend to originate in the Gulf of Mexico or the Caribbean Sea; the storms that strike later in the season (August, September, and October) are likely to form in the eastern North Atlantic Ocean off the coast of Africa.

Although meteorologists are still seeking answers to questions about the causes and mechanics of hurricanes, the basic model of what happens is known. During the summer the surface waters off West Africa heat up to at least 79°F. Evaporation produces a layer of warm, moist air over the ocean. This moist air is trapped beneath warm air coming from the African continent, but some of it is drawn upward. As the moist air rises, it cools and condenses, releasing heat, which in turn warms the surrounding air and causes it to rise. As the mass of rising air increases in size, a low-pressure area forms (tropical depression), and warm easterly winds rush in to replace the rising air. The effect of the earth's rotation deflects that air flow, and the counterclockwise-rotating air mass begins to take on the familiar spiral shape of a hurricane. Air forced to the middle of the spiral can only move upward, producing a chimneylike column of rising air—the "eye" of the storm.

In sum, a heat-engine effect evolves, with rising moist air cooling and condensing, releasing heat that causes more air to rise, bringing more air rushing in over the sea, an endless source of moisture. Heavy rainfall characterizes the edges of the cloud mass, and when the sustained wind velocities reach 74 miles per hour the storm is classed as a hurricane. The strongest winds of a hurricane may exceed 200 miles per hour. The maximum wind speeds of the largest storms to hit coastal areas are generally unknown because the wind-measuring instruments were blown away.

Once formed, the hurricane mass begins to track into higher latitudes and may continue to grow in size and strength. The velocity of this tracking movement can vary from nearly stationary to greater than 60 miles per hour. If you consider that the diameter of a hurricane

3.1. Satellite image of Hurricane Hugo making landfall near Charleston on the night of September 21, 1989. Photograph courtesy of NOAA.

Table 3.1 The Saffir/Simpson Hurricane Scale

	Category				
	1	2	3	4	5
Central Pressure, inches of mercury (millibars)	≥28.94 (≥980)	28.91–28.50 (979–965)	28.47–27.91 (964–946)	27.88–27.17 (945–921)	<27.17 (<920)
Winds, miles per hour (meters/second)	74–96 (32–42)	96–110 (42–49)	111–130 (50–57)	131–155 (58–68)	>155 (>69)
Surge, feet (meters)	4–5 (1.32)	6–8 (2.13)	9–12 (3.20)	13–18 (4.57)	>18 (>5.49)
Damage	Minimal	Moderate	Extensive	Extreme	Catastrophic

Source: Developed by H. Saffir and R. H. Simpson; "The Hurricane Disaster Potential Scale" (appendix D, ref. 12).

ranges from 60 to 1,000 miles, and that gale-force winds may extend over most of this area, the total energy released over the thousands of square miles covered by the storm is almost beyond comprehension. No ship or seawall, no cottage, condominium, or other static structure is immune to the impact of such forces. For a hurricane making landfall in South Carolina, these forces will be at their maximum in the area to the right (north or east) of the eye, but the entire landfall area will experience the severity of the storm. If the hurricane comes on a high tide, especially a spring tide, the effects of storm-surge flooding, waves, and overwash will be magnified.

Hurricane Probability and Rank

The probability that a hurricane will occur in any one year along any 50-mile segment of the South Carolina coast ranges from 5 to 8 percent; the likelihood of a great hurricane occurring is 1 percent. Such low numbers may encourage a false sense of security, but hurricane history tells us that such a storm is almost a certainty at least once during the life of a structure. Furthermore, the fact that a great hurricane occurs one year does not in any way reduce the likelihood that a similar storm will strike the next year.

Along the Georgetown County coast there is

a legend that the Grey Man walks the beaches on the eve of a hurricane (appendix D, refs. 2, 3). Pawleys Island claims to be the home of this well-known South Carolina ghost, although he has been seen as far south as Isle of Palms. He is variously reported giving warning, gesturing, walking the beach, or making last-minute hurricane preparations on ghostly structures. Unfortunately, this prophet of disaster never comes in time to spread the alarm so that coastal residents can evacuate. Today, the hurricane watchers of the National Oceanic and Atmospheric Administration track hurricanes and provide advance warning so that threatened coastal areas can be evacuated. Still, as little as 9 to 12 hours of advance warning may be all that is possible, given the unpredictable turns a hurricane may take. That is not much better than the Grey Man's prophecy, but we also benefit from radio, television, and having the means to evacuate quickly—if we take the warning seriously.

The only information available to us about the strength of particular past hurricanes comes from sketchy newspaper accounts describing the damage and the number of lives lost or estimating the losses in dollar amounts. Major storms such as the ones in 1752, 1804, 1813, 1822, 1854, and 1893 were all characterized as "the worst ever" or "greater than" some previous "worst" storm. The hurricanes were not, in fact, increasing in intensity, but each storm followed a somewhat different path, so that the full fury was felt at Beaufort

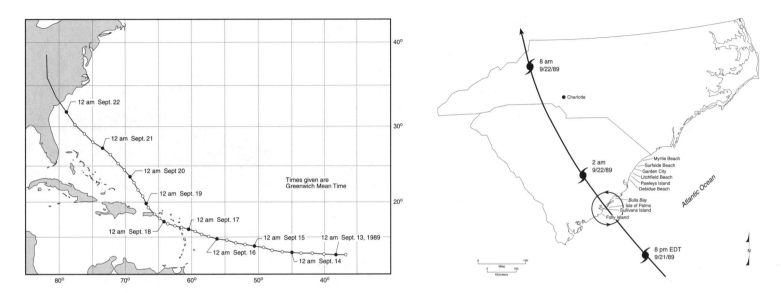

3.2. Track of Hurricane Hugo. (A) Left: Hugo moves across the Atlantic Ocean, September 10–22, 1989. (B) Right: Approach, landfall, and inland track of Hugo across South Carolina and North Carolina.

in one storm and at Charleston or Georgetown in another. Modern storms are sometimes compared in terms of dollar losses, but that figure reflects the nature of the development that was struck rather than the strength of the hurricane. Likewise, the smaller storms of less than a century ago were more deadly than the largest of today's storms. Advance warning, efficient evacuation, and safer construction today should keep casualty rates low even in a major hurricane. But unsafe development, al-

lowing population growth to exceed the capacity for safe evacuation, and complacency on the part of coastal residents could reverse this trend, with shocking results.

The National Weather Service has adopted the Saffir/Simpson scale (table 3.1) for communicating the strength of a hurricane to public safety officials of communities potentially in the storm's path. The scale ranks a storm on three variables: wind velocity, storm surge, and barometric pressure. Although hurricane paths are still unpredictable, the scale communicates quickly the nature of the storm; that is, what to expect in terms of wind, waves, and flooding.

Do not be misled by such scales, however. A

hurricane is a hurricane. The scale simply defines how bad is bad. When the word comes to evacuate, do it. The wind velocity may change or the configuration of the coast may amplify the storm-surge level, so the category rank can change. Don't gamble with your life or the lives of others. Appendix A provides a checklist of things to do when a hurricane threatens.

Hugo: New Lessons in Coastal Processes

In the early days of September 1989, a U.S. Weather Service satellite detected a cluster of thunderstorms moving westward off the coast of Africa. Over the course of the following week this disturbance grew stronger and more

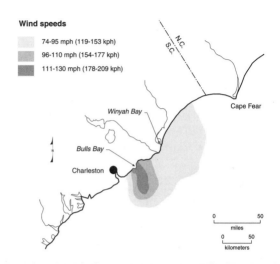

Wind speeds

74-95 mph (119-153 kph)

96-110 mph (154-177 kph)

111-130 mph (178-209 kph)

N.C.
S.C.

Cape Fear

Winyah Bay

Bulls Bay

Charleston

0   50
miles
0   50
kilometers

3.3. Wind speeds recorded during landfall of Hurricane Hugo. Tropical storm-force winds extend well beyond the shaded area.

organized. When its winds reached tropical storm level, it was christened Hugo. On September 13, Hugo became a hurricane. A week was to pass before Hugo made landfall in South Carolina.

It is not possible to predict a hurricane's path. But, sandwiched between stable air masses, Hugo's track across the Atlantic toward the eastern seaboard was steady and ominous (fig. 3.2). By mid-September meteorologists were clearly worried. Hugo's wind speeds had reached 150 miles per hour. Someplace was going to get hit very hard.

Hugo's first victims were Guadeloupe, Dominica, and their neighbors in the Lesser

Antilles. Hugo next moved on to ravage the Virgin Islands. On September 18, more than 90 percent of the buildings on the island of St. Croix were destroyed or damaged. Much of the vegetation of the island was stripped bare. Then Hugo hit the northeastern corner of Puerto Rico. Serious damage there was limited mainly to the coastal areas due to the low storm surge and the steep coast. Walkways, roads, seawalls, and structures in the low areas were destroyed by waves 12 to 20 feet tall. Even though Hugo was a relatively dry storm, its associated rainfall triggered hundreds of landslides in the island's interior. Hugo had completed its warm-up.

As Hugo crossed the Gulf Stream on September 21, reconnaissance aircraft reported that pressures in the eye were falling. The maximum sustained winds were up to 138 miles per hour. Hugo was a category 4 storm. A hurricane warning was issued at 6:00 A.M. for the coast from south-central Florida to Cape Lookout, North Carolina. Barrier islands and low coastal areas were evacuated. Hugo's eye was 40 miles wide.

Gone with the Wind

Mayor Joseph Riley called it the greatest disaster that Charleston had ever faced. Certainly Charleston had seen other hurricanes. The city had survived floods and great fires. The harbor's shores had endured British naval assaults and a cruel siege by the Union Army and

3.4. Top: The Ben Sawyer Bridge, a swivel-type bridge to Sullivans Island, was blown off its foundation by Hurricane Hugo's fierce winds.

Navy during the War Between the States. Charleston had even been the epicenter of the greatest earthquake ever recorded in the southeastern United States. But Mayor Riley was correct. Charleston and its suburbs were about to be overrun by a tremendous natural disaster—a great hurricane.

As the eye wall of Hugo moved over land, the wind speed increased rapidly (fig. 3.3). Gusts of 125 miles per hour were recorded by navy ships in Charleston Harbor, and the Charleston Naval Base weather station reported 137 miles per hour. Maximum sustained winds were estimated at 135 miles per hour at Bulls Bay, upcoast from Charleston. Hurricane-force winds extended approximately 100 miles upcoast toward the northeast and 50 miles to the southwest.

3.5. Bottom: Direct wave damage from Hugo in the Grand Strand area.

The most common, and often the most costly, of storm hazards causing damage is direct wind impact on buildings and other structures. Strong winds also cause tremendous damage to vegetation of all kinds and are responsible for transporting sediment onto and off of the islands.

As Hugo came ashore, the works of man began to disintegrate. The Ben Sawyer Bridge, a swing bridge over the Intracoastal Waterway, broke loose just before midnight and began to spin. After a few revolutions the bridge collapsed and turned askew, pointing toward the sky (fig. 3.4). Boats in the City Marina off the lower peninsula were lifted as if en masse and dumped in a heap on Lockwood Boulevard. As water rose in the Holy City, slate roofing tiles brought from Wales as the ballast for tall ships scattered in the wind.

Few buildings on the barrier islands escaped significant damage, which ranged from loss of roof and wall coverings to complete destruction. The damage was greatest in the areas that received the highest winds, of course, but the harm done to many hundreds of buildings did not correlate with the worst storm conditions. Instead, this needless damage was attributed to an inappropriate building code, difficulties in code enforcement, and the unusually long interval since the last major hurricane (appendix D, ref. 160).

Pine forests are the terminal vegetation for the sandy soils of the coastal zone. More than 90 percent of these pines were snapped off by Hugo at a strangely consistent height above the ground. Eighty-six percent of the more brittle hardwoods were either uprooted or broken to pieces. Approximately 1.8 million hectares of South Carolina forest land were damaged by wind and water. By comparison, the eruption of Mount Saint Helens affected 60,750 hectares and the Yellowstone fires of 1988 burned 400,000 hectares (appendix D, ref. 34)—3 percent and 22 percent, respectively, of the area destroyed by Hugo.

Waves

Buildings are damaged both by direct wave attack (fig. 3.5) and by pummeling (ramrodding) by floating debris. Of course, waves are also responsible for shoreline erosion, dune erosion, overwash, and destruction of vegetation.

Storm surges expose much wider areas than usual to wave attack. Breaking waves move across the surge zone, and Hugo's combination of waves and flooding planed off dunes as well as lifting and floating houses off their foundations (fig. 3.6).

Waves from even a small hurricane can be highly erosive if the storm remains poised offshore for an extended time. In this regard Hugo was unusual because it never faltered. It approached the shore at 30 miles per hour and churned over the coastal zone and on into the interior. The damage done by the other storm components of Hugo far exceeded the battering of ocean waves on the shore.

3.6. House moved off its foundation by waves and storm surge. Floating buildings become battering rams that plow into adjacent buildings, increasing the storm-related damage and destruction.

Storm Surge

Storm surge is a local rise in sea level caused by hurricane winds, circulation, very low barometric pressure, and offshore water pushed into the shallower nearshore. Storm surge causes flooding and also extends the zone of wave impact inland. The results are structures floating off their foundations (fig. 3.6), debris floating inland, ramrodding by such debris, and general water damage.

The area just outside and to the right of the storm's eye carries the greatest energy and strongest winds. Storm surge also is usually greatest in this area, which can extend tens of miles out from the center of the eye, depending on the size of the storm (fig. 3.7). Flood levels rise quickly where this margin meets the land, giving rise to such misnomers as "tidal wave" and the inevitable "wall of water." Though scientifically inaccurate, these descriptions seemed appropriate for Hugo's arrival in the area around Bulls Bay.

The storm surge was already at work well before Hugo made landfall. Rivers and estuaries had filled and overflowed. From Kiawah Island to the North Carolina border the barrier islands were completely under water. Thirty-odd miles upcoast from Charleston the counterclockwise winds piled up a surge more than 20 feet high. The towns around Bulls Bay were devastated. Shrimp boats were strewn randomly throughout the residential area of McClellanville. The water level in Awendaw reached nearly 19 feet. In the small community of Romain Retreat, all that remained of most houses was debris around the driveways.

Farther north from Bulls Bay the storm surge quickly overtopped the seawalls and dunes of the Grand Strand, actually reducing the potential shoreline erosion. Storm surges varied little here, ranging from approximately 12.0 to 13.5 feet throughout the Strand. Houses on the spit of Pawleys Island were washed into the marsh. Many houses from

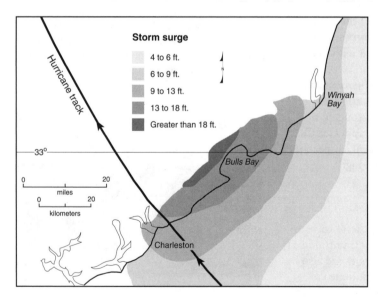

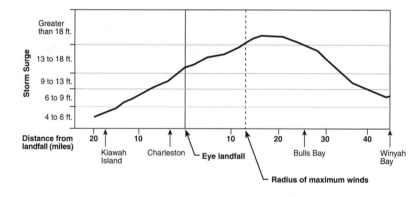

3.7. Characterizing the storm surge associated with Hurricane Hugo. (A) Left: Surge map shows regional increases in water levels from Charleston, through Bulls Bay, to Winyah Bay. (B) Right: Surge cross section showing the typical mound shape of the water mass that moves with the storm.

Garden City north into North Carolina were either destroyed in place or washed landward into other structures. The 18- to 20-foot dune at Litchfield, built following Hurricane Hazel in the 1950s, remained intact and did provide some protection from the surge. But flooding continued from the landward side. And of course there were the winds.

Not surprisingly, the worst damage from Hugo's surge occurred in Charleston County, where the eye came ashore. South of the eye the surge level exceeded 10 feet. With its starved beaches and tired seawalls, Folly Beach was a bone to be chewed. Even greater flooding occurred on Sullivans Island and Isle of Palms. Here, where the right side of the eye made landfall, water levels rose 12 to 14 feet. Wide beaches and good setbacks could do little to mitigate the intensity of this flooding. No place escaped damage. In many places the destruction was complete.

Coastal Erosion

Coastal erosion is a result rather than a process, but it is a measure of an island's response to coastal processes. Hugo produced some of the greatest short-term erosion of modern times. Kiawah Island experienced major erosion at its north end. At North Myrtle Beach massive revetments were overtopped and destroyed. Between these extremes the nature and degree of beach erosion varied with the shoreline. The damage was not limited to

South Carolina. The beaches of Brunswick County, North Carolina, a hundred miles away from the track of Hugo, also experienced major erosion.

Fortunately, Folly Beach was spared the worst of the storm surge. This highly eroded island has a very narrow beach and no protective dunes. Nevertheless, seawall destruction was greater than 95 percent, and erosional escarpments were found underneath and landward of houses—those that remained. A large number of houses were driven landward, blocking streets or crashing into other houses. Some had a stranger fate. Once lifted from their foundations, they were carried offshore by the returning storm-surge ebb. The remnant debris of these structures lies buried in sand about 200 yards offshore.

Beach erosion was at first difficult to see amid the immense destruction of the beachfront property on Sullivans Island and Isle of Palms. For the most part these islands have wide beaches and an abundance of sediment in the nearshore area. Hugo significantly changed their beach profiles, however, leaving behind narrower beaches with a lowered and flattened slope. A new shoreline had been configured. The average erosion was 12.5 cubic yards of sand from each linear foot of the beach (appendix D, ref. 34). Most of this sand was not carried far offshore, however, and it migrated back onshore over the following weeks and months. As bad as it was, an observer would be hard put to identify remnants

of Hugo's beach destruction here today. Buildings were rebuilt, more buildings were added, beaches were nourished, newly carved inlets were filled, and the debris was cleared away. A remarkable recovery, but also a stage reset for the next performance.

Along the Grand Strand, Hurricane Hugo caused landward recession of the upper beach by 80 to 150 feet (appendix D, ref. 29). Beach erosion extended up to elevations of 13 to 14 feet above mean sea level. Sand dunes and seawalls were no help against this storm, either for the beaches or for the structures behind them.

Storm-Surge Ebb

The storm surge from a major hurricane may push tremendous volumes of water over an area of many tens or even hundreds of square miles. If the surge comes on a spring tide, the flooding is magnified. And the same catchment basin may be receiving tons of rainfall as well, raising the water level from freshwater runoff. When the storm passes or simply lessens, gravity forces this superelevated water to drain back into the ocean. In spots where the flow is confined by natural landforms or buildings and infrastructure, strong currents develop, leading to increased erosion and damage. The hurricane's track or the shoreline configuration may be such that the winds, now blowing offshore, actually blow the water back to the sea, increasing the ebb-flow effect. The flow scours new channels, changes the shape and orientation of existing channels, and cuts new inlets from the back side of the islands, as occurred on the southern end of Pawleys Island (see chapter 6). On barrier islands this force can carry untold amounts of debris offshore, including houses. The water draining across the land and down the beachface carves ebb-scour channels (fig. 3.8).

A post-Hugo study on Folly Beach (appendix D, ref. 37) identified 30 ebb-scour channels, an average of 3 per city block. They varied in dimension but were up to several hundred feet long, equally wide, and up to 10 feet deep. Beach erosion from these events may be small compared with other kinds of hurricane damage; however, serious damage occurs to dunes, structures, and roads caught in the way of an ebb-scour flow (fig. 3.9).

Two precautions can greatly lessen the danger posed from ebb-scouring. Ebb-scour channels left behind by Hugo were most often associated with unvegetated ground where sand was mobilized by even a small amount of flowing water. Often nothing more than the presence of lawn grass prevented the growth of an ebb-scour channel. The second factor that controlled or contributed to these channels was failed or weakened erosion control devices. If a coastal property is fronted by a seawall, that structure must have a high degree of integrity to prevent ebb-scours. Also, healthy vegetation is imperative for dunes and grounds to survive a hurricane.

Tornadoes

Hurricanes sometimes generate tornadoes when they make landfall, usually in the area to the right and front of the eye. Early reports following Hurricane Hugo claimed that as many as 250 tornadoes had been spawned by the storm, and officials used this preliminary number for weeks afterward. A closer look by experts reduced the number of confirmed tornadoes. And what was the final number of *confirmed* hurricane-spawned tornadoes from Hugo? None! According to *Impacts of Hurricane Hugo: September 10–22, 1989,* edited by Charles W. Finkl and Orrin H. Pilkey (appendix D, ref. 25), there were no confirmed tornadoes generated by Hugo. Several probably did occur inland, in the areas of Florence, Sumter, and Georgetown Counties in South Carolina, and in Cherokee County and Hickory, North Carolina. But the widespread damage caused by the hurricane winds prevented tornado signatures from showing up on aerial surveys. Why the confusion? Hurricanes do spin off tornadoes, but the tornadoes cannot endure

3.8. Ebb-scour channels, outlined in black, formed on Sullivans Island along shore-perpendicular lines where vegetation was absent and walls failed.

3.9. Ebb-scour channel running seaward on Folly Beach. Erosion occurred because the return ebb flow was concentrated between the houses, along an unvegetated beach access path into a vacant beachfront lot (submerged in background). Photograph by Ralph Willoughby, South Carolina Geological Survey.

for long in the phenomenal winds associated with large hurricanes. They may form and last for a brief time, causing serious wind damage, but they are quickly destroyed by the violent wind forces that created them. Because twisters' wind velocities may not greatly exceed those of the hurricane that spawned them, their wind damage is impossible to separate from hurricane destruction.

New research is disclosing many interesting facts about the incredible winds generated by hurricanes. Experts and Weather Service personnel now use the terms *microburst* and *downshear* to describe these ephemeral events, which can exceed tornado-like speeds, greater than 400 miles per hour. Also, they are commonly oriented horizontally, like a tornado turned on its side. These violent, short-lived winds can circulate either clockwise or counterclockwise, and are the only winds known to be capable of circulating in either direction in the same hemisphere. Such localized winds may account for some of the more unusual kinds of hurricane destruction, such as part of a house and rooftop destroyed and the rest apparently undisturbed. Although microbursts

and downshear may be rare amid the general mayhem of a hurricane, one should take all possible construction steps to ensure the integrity of structures that, sooner or later, will be in the path of a killer storm.

Northeasters and Winter Storms

On December 1, 1986, a northeaster descended the Atlantic seaboard and had its way

with the South Carolina coast (fig. 3.10). Looking back after Hugo, the damage from that event seems insignificant. At the time, however, it was devastating: the storm destroyed many dozens of seawalls in South Carolina alone and caused erosion on beaches from the Grand Strand to Hilton Head. Ashley Avenue, at the Folly Island washout, was breached—not for the first time. Nor would it be the last.

3.10. A small northeaster impacting Folly Beach, December 1, 1986. Photograph courtesy of the *Charleston Post-Courier*.

call the Ash Wednesday Storm, the northeaster of 1962 that was one of the most widespread and destructive winter storms of all time. From the Carolinas to New York, hundreds of houses were damaged and destroyed, new inlets were breached, seawalls and groins were destroyed, and ships ran aground.

The Nature of Northeasters

Northeasters are far more common than hurricanes. There are about 30 per year, although only a few affect South Carolina. They are also much larger and longer lasting than hurricanes. Hurricanes are typically 300 to 400 miles in diameter, with their greatest winds concentrated around an eye wall 50 to 60 miles across. A hurricane's exposure to any given area of the coast is usually measured in hours. Winter storms are not so concentrated. A northeaster is commonly spread out over a thousand miles and usually hugs the Atlantic seaboard for days.

Northeasters typically form as low-pressure cells over coastal areas, where there is a large difference between air temperatures (cold over the land and relatively warm over the water; see appendix D, ref. 43). The waters off the Carolinas are a major breeding ground for big northeasters because this area lies close to the

Only one month later, on January 1, 1987, it happened again. This time the northeaster (also called "nor'easter," or simply "winter storm") was substantially larger, tearing out more seawalls and eroding the beaches even more than the storm of the previous month. The washout was repaired again, and made ready for Hugo. That storm has since been called the New Year's Day Storm, in the tradition of naming significant storms after the holi-

days or special occasions on which they seem to occur.

Perhaps we remember an event better, for better or worse, if it is associated with a celebration, religious holiday, or anniversary. But really, it is the loss or tragedy that makes the day memorable. Many coastal residents remember the New Year's Day Storm, just as many recall the Thanksgiving Day Storm and the Halloween nor'easter of '91. Many still re-

Table 3.2   The Dolan/Davis Northeaster Intensity Scale

	Storm Class				
	1 (weak)	2 (moderate)	3 (significant)	4 (severe)	5 (extreme)
Beach erosion	Minor changes	Modest: confined to lower beach	Extends across entire beach	Severe beach erosion and recession	Extreme beach erosion (up to 50 m in places)
Beach recovery	Full and usually immediate	Full	Usually recovery over considerable time (months)	Recovery seldom total	Permanent and clearly noticeable changes
Dune erosion	None	None	Can be significant	Severe dune erosion or destruction	Dunes destroyed over extensive areas
Dune breaching	No	No	No	Where beach is narrow	Widespread
Overwash	No	No	On low-profile beaches	On low-profile beaches	Massive in sheets and channels
Inlet formation	No	No	No	Occasionally	Common
Property damage	No	Minor, local	Loss of many structures at local scale	Losses of structures at community level	Extensive regional scale: millions of dollars

Source: "Nor'easters" (appendix D, ref. 42).

Hugo and Associates   39

winter track of the polar jet stream. The influence of fast upper-level winds on the formation of northeasters helps distinguish them from hurricanes.

The northeaster develops a counterclockwise-rotating air circulation, generating winds that blow onto the east coast from out of the northeast when the storm center is out over the Atlantic, hence the name. Most often these cells track up the Atlantic seaboard, gaining strength from the warmer ocean waters. A major factor in the development of destructive winter storms is the required presence of a strong, stable high-pressure system over eastern Canada. This prevents the storm from moving quickly to the north or northeast and ensures that it will remain off the Atlantic coast for a long period, often several days. Storm surge and waves are usually the most destructive coastal processes during a northeaster.

A classification system comparable to the Saffir/Simpson scale has been developed to rate the strength of northeasters (see table 3.2; appendix D, ref. 42). The Dolan/Davis system uses wave height and storm duration to determine five categories of winter storms. Each successive category is associated with a greater degree of damage to coastal structures and landforms. It is practical and easy to use; one computes a storm's intensity by noting the number of hours the storm's waves exceed 5 feet. The system therefore gives a measure of a storm's overall intensity and can be used to predict future property damage. Advance warning of the intensity of a northeaster is of real value to residents and emergency planners. Evacuation is usually not a regional problem with northeasters, but island fronts and low-lying areas are subject to storm-surge flooding, and local evacuation may be necessary. As the Atlantic Ocean shores become more densely populated, greater lead time for preparation and evacuation will become more necessary to protect human lives and property.

Increasing development puts more and more people and property at risk from hurricanes and northeasters. How should coastal residents respond to the increased risk? Use the information presented in chapters 5 and 6 to help you select a safe site to build or buy a house. Build a house that will stand up to hurricane forces (see chapter 8). If your house is already built, secure it according to the guidelines presented in chapter 8.

4 Responding to Coastal Change: Going with the Flow

The fundamental problem with living at the shore, or even in the coastal zone, is that all of our static, immobile construction (buildings, roads, bridges, and utility supply lines) is placed in a dynamic zone. The shoreline, inlets, dune fields, overwash terraces, marshes, and maritime forests shift landward as well as laterally in response to ongoing changes in the coastal zone. Changes in the levels of the sea and the land, changes in storm frequency and wave regime, change in the patterns of currents, change in the offshore topography, change in sand supply, changes in the controls on vegetation growth, and changes we are only just beginning to recognize (e.g., climatic change and changes in water quality due to pollution)—change is the rule, especially for barrier islands. The history of South Carolina's coast is one of change, both natural and socioeconomic.

How have we responded to the rising sea level and barrier island migration, that is, to shoreline retreat? Until recently, the most common response was to try and stop it—to stabilize the beach or inlet exactly where it was; to prevent erosion of the land at the back of the beach; to make nature static and immobile like our buildings and roads. At the same time we have found it desirable to change the natural texture of the barrier islands. We level dunes, block overwash, reduce the cover of natural vegetation, replace native species with exotic ones, dig finger canals and dredge channels—we replace the "flow" of nature with the grid of our suburban houses, streets, and services. The price for this approach has been and will continue to be enormous. Hurricane Hugo's record cost will be eclipsed sooner than we think.

Shoreline Engineering

The history of human "progress" has been to claim land, to build on it (develop it), and then to defend it. Over the centuries, we have progressed from defending developed coastal land from invading nations to defending it against the invading sea. The job has moved from the purview of armies and navies to that of the coastal engineer. In the United States, the major coastal engineering specialists are still linked to "defense" through the U.S. Army Corps of Engineers, often known simply as the Corps.

Shoreline engineering is a general term that refers to methods of changing or altering the natural shoreline system in order to stabilize it. Stabilization methods range from the simple to the complex—from planting dune grass to constructing large seawalls using draglines, cranes, and bulldozers. The benefits of such methods are usually short-lived. Locally, shoreline engineering may actually cause shoreline retreat, as evidenced by the loss of beach in front of the North Forest Beach seawall on Hilton Head Island and the high rate of erosion on Folly Beach near the Charleston jetties. The beach erosion resulting from coastal engineering may be greater and more spectacular than any caused by nature. There are, of course, a few situations in which stabilization is undeniably an economic necessity. The channels leading to the major state ports of Charleston, Georgetown, and Port Royal, for instance, must be maintained.

The economic price of stabilizing a beach can be stiff indeed. In 1980, the initial cost of the proposed erosion control projects along the South Carolina coast was estimated at $38 million. The projected annual maintenance costs for those projects was in excess of $5 million, and future inflation would result in higher figures. By the 1990s, the estimated cost for the state's beach nourishment to the end of the century alone was $65 million.

Public awareness of the magnitude of the erosion problem grew, and by the mid-1980s the state was moving toward better shoreline management. In 1990 the final Beachfront Management Act became effective (see chapter 10 and appendix C). Under this forward-looking law, new hard structures such as seawalls and revetments are not permitted. But the die had already been cast for much of South Carolina's developed coast; of approximately 90 miles of developed shoreline, more than 18 miles was totally stabilized, nearly 5 miles was mostly stabilized, and more than 4 miles was partially stabilized (appendix D, ref. 91). Folly Beach had gone from 28 percent stabilized to 60 percent stabilized in just eight years.

Fortunately, the beaches of South Carolina

4.1. Beach nourishment at Wild Dunes on Isle of Palms. A mixture of water and sand is pumped in through the pipe shown in the foreground. Note the condominiums in left background fronted by a seawall. Photo courtesy of OCRM.

Table 4.1   The Advantages and Disadvantages of Beach Replenishment

Advantages:
The beach is widened
Temporary protection of property
Storm protection
Maintains the recreational value of the beach
No negative impact on downdrift beaches
  (becomes a better sand source)
Looks better (aesthetics)

Disadvantages:
Temporary, must be renourished
High cost (increases with each new nourishment)
Decreasing sand supply
Possible damage to marine organisms from
  turbidity
Possible unforseen environmental impact
Offshore dredge hole may create erosive wave
  refraction pattern, may be a sink for mud
Creates a false sense of security, spurring
  additional development in high-hazard zone

have fewer engineering mistakes on them than certain beaches in other states (New Jersey and the southern part of Florida, for example). Despite this, and even though new hard structures are banned, the existing structures will continue to cause significant shoreline changes for decades to come. South Carolinians, whether beachfront property owners, beach users, or simply taxpayers, should be aware of the effects of shoreline engineering structures and their long-term costs. If the fundamental principles are forgotten or ignored, the pressure from front-beach property owners may result in the law being relaxed or changed, and these beach-destructive devices may be allowed again.

There are three major types of shoreline stabilization: beach replenishment, groins and jetties, and seawalls and breakwaters. These are discussed below in order of their decreasing environmental compatibility.

Beach Replenishment, Nourishment, and Renourishment (Soft Stabilization)

South Carolina has set in motion a policy of retreat from rising sea levels, except where beach replenishment can be economically justified. By doing so, the state has opted for what is probably the most gentle approach to beach repair. Replenishment consists of pumping or trucking sand onto the beach and building up the dunes and upper beach that have been lost (fig. 4.1; table 4.1). Sufficient money is almost never available to replenish the entire beach, which extends out into the water to a depth of about 30 feet. Thus, only the upper beach is covered with new sand, in effect creating a steeper beach.

The steepened beach quickly becomes less steep, because a replenished beach erodes very rapidly. Depending on a number of factors, such as the grain size of the sand and the storm frequency, a replenished beach will erode as much as 10 times faster than a natural beach. Seabrook Island, Folly Beach, Hilton Head Island, Hunting Island, Debordieu, and a few other South Carolina beach replenishment projects have not held up well in terms of longevity. Even more fundamental than beach steepening by renourishment as a cause of erosion is the problem of the rise in sea level.

The cheapest and least environmentally dangerous sand source for beach replenishment is the mainland; for most projects, however, the sand is pumped either from the back side of the island, from a pit on the island, or from an offshore site on the continental shelf. Sand from the back side tends to be too fine, and it quickly washes off the beach, as illustrated by the rapid erosion of Folly Beach's 1993 replenishment. On the other hand, if the nourishment material is too coarse (e.g., if it consists of broken oyster shells), the recreational value of a bathing beach may be lost. Dredging in back of the island also disturbs the ecosystem, and the resulting cavity may change wave and current patterns, which can accelerate erosion around the borrow site. Finally, the mud released by dredging is detrimental to filter-feeding organisms such as clams and coral.

As sand supplies have become scarcer and replenishment demands larger, attention has turned to offshore sand; but environmental problems make this a costly alternative. The borrow hole becomes a mud trap for future resuspension. Furthermore, wave patterns are altered when a hole is dug on the shelf, and a change in the wave energy distribution on the adjacent shoreline can cause increased erosion. One additional point: use of a certified ocean-going dredge normally raises the price of a beach replenishment project by at least $1 million.

Beach replenishment also can create a false sense of security because it often conceals past storm damage. In addition, the value of property behind a nourished beach may increase, and this spurs even more development, including condominiums. The increased development density further increases the demand (and political pressure) from the shore community for the next round of beach replenishment. This has been the fate of several Grand Strand communities. Beaches cannot be replenished *ad infinitum*. Sand supplies become exhausted, more distant, and more costly. The natural sand supply diminishes as the stabilized island, held in place where we think it should be, becomes more and more out of equilibrium with the rising sea level. Inevitably, the island community will want more drastic and costly stabilization measures. This pressure creates long-term political disequilibrium.

The most celebrated beach replenishment of recent years was the Miami Beach project, which replaced 15 miles of beach at a cost of $65 million. In general, the cost of beach replenishment is around $2 million per mile in the United States, but the price is going up and can vary widely. Consider that virtually hundreds of miles of American shoreline have buildings crowded close to the beach, and that all of these communities will soon be "in danger" from shoreline erosion because the sea level is rising. If the majority of beach communities seek to stabilize their shorelines, the cost to the taxpayer—local and federal—will be tremendous. A taxpayers' rebellion is already brewing, as reflected by passage of barrier island bills and flood insurance reforms (see chapter 10). The Coastal Barrier Resources Act prevents future expenditure of federal funds (federal subsidies) on presently undevel-

4.2. Photos taken in 1993 and 1995 of the replenished beach just north of the Folly Beach Holiday Inn. Note the dramatic loss of sand in the first two years of the projected eight-year beach. (A) Left: May 1993: just a few months after replenishment the new beach was already disappearing. Note that the top of the groin is only slightly exposed. (B) Right: The same groin in 1995: much of the sand is gone from the recreational beach, and the previously buried groin is now exposed.

oped barrier islands. Support for beach replenishment most likely will become more and more difficult to obtain. South Carolina's recreational community beaches increasingly are becoming artificial beaches. Grand Strand

beaches, Edisto Island, Folly Beach, Hilton Head, Debordieu, and Seabrook have all required nourishment to hold the line: to maintain the now artificial position of a buffer beach whose natural life is limited in time and space.

Beach Replenishment: The Story Behind the Story

By law, replenished beaches must be justified as storm protection for the community if they are to receive federal funding. Creating a new recreational beach is not an acceptable justification for constructing a replenished beach. Of course, most communities want replenished

beaches because a wide and handsome new beach improves both the economy and the quality of life.

As a major player in beach replenishment in the United States, the U.S. Army Corps of Engineers has a vested interest in promoting replenished beaches. The story of how the Corps replenished Folly Beach should be required reading for any community considering replenishment. The Corps convinced the community of Folly Beach that beach replenishment was cost-effective and would restore the heavily eroded beach. The Charleston District of the Corps said the beach would not need new sand for eight years (i.e., it predicted an eight-year nourishment interval). In less than two years,

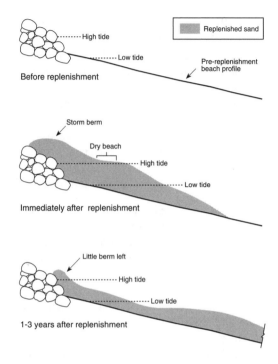

Before replenishment

Replenished sand

High tide

Low tide

Pre-replenishment
beach profile

Storm berm

Dry beach

High tide

Low tide

Immediately after replenishment

Little berm left

High tide

Low tide

1-3 years after replenishment

4.3. Evolution of a replenished beach. Replenishment reestablishes the recreational beach, but only for the short term. Both the economic value of the beach and the protective storm berm fail to persist over the predicted life span.

however, the wide replenished beach was gone, along with much of the Folly Beach shoreline (fig. 4.2). This occurred over a time when no significant storm affected the coast, and far more quickly than local citizens had been led to believe it would. In response to criticism, district spokespersons claimed that there was nothing to worry about because all the sand was still "in the system," just offshore. The flawed implication was that if sand can be accounted for by sophisticated survey-

ing techniques, it can still protect the community from storms. This has become the reply of choice from Corps districts whose beaches disappear faster than anticipated.

You've heard of the type of statistics that can make a journeyman baseball player sound like a Hall of Famer? Sure he bats over .400— on grass fields, at night, facing left-handers, during weekend games, with nobody on and nobody out. Similar logic is used to make beach replenishment projects appear successful. One year after the Folly Beach replenishment sand was pumped in, the Corps of Engineers said that 95 percent of the beach could still be accounted for. Two years after emplacement, 89 percent of the sand was still "in the system." Beachgoers are unhappy and unimpressed with *underwater* sand. Underwater sand offers no space for sunbathing or playing beach volleyball (though surfing may improve with erosion). The narrow to nonexistent swimming beach is not what the city of Folly Beach hoped for, and the storm berm no longer offers much storm protection. Furthermore, offshore sand does not offer much storm protection, despite claims to the contrary. This was the state of most of Folly's beach approximately two years after completion of the re-

plenishment project, with six years to go until the projected "next needed" nourishment.

Figure 4.3 illustrates what happened at Folly Beach. Immediately after replenishment there was both a protective berm and a wider dry beach (there was a recreational beach at high tide). The beach profile also was steepened somewhat, however, and wave energy began to redistribute the sand out into the subaqueous portion of the beach. The sand moved out beyond the surf zone is still "in the system," but it no longer offers much resistance to storm waves, especially if a storm surge occurs, which almost always is the case in a major storm. The protective storm berm is diminished, and the high-tide dry beach is either very narrow or absent.

More replenishments are on the drawing board in South Carolina and elsewhere, and community residents should have a clear and honest understanding of what lies in store. The projected nourishment interval for North Myrtle Beach and Myrtle Beach is 10 years. The assumed or projected rate of erosion of the new beach is such that it should last 36 years before the last grain of sand disappears. We predict that 3 to 5 years is a more likely life span. For Garden City, the Corps says sand will have to be pumped in every 6 years. These replenished beaches are unlikely to last nearly as long as the Corps has predicted, however, and the unlikelihood grows with each passing storm. The Corps will tell city officials that almost all of the sand is "accounted for" and not to worry. We believe underwater sand doesn't protect and shouldn't count, and that people should tell this to the U.S. Army Corps of Engineers.

Questions to Ask If Beach Nourishment Is Proposed

When a community is considering some form of beach maintenance shoreline engineering, it is almost invariably in an atmosphere of crisis. Buildings and commercial interests are threatened, time is short, an expert is brought in, and a solution is proposed. Under such circumstances the right questions are sometimes not asked. The following is a list of questions you should bring up if you find yourself a member of such a community.

1. How long will the proposed solution protect the recreational beach as well as property? 10 years? 20 years? 30 years? 50 years? What is the basis for the prediction—a mathematical model or experience on similar or nearby beaches?
2. How much will maintenance of the artificial beach cost in 10 years? 20 years? 30 years? 50 years? If the cost is projected to be less than $1 million per mile per year, be wary of an overly optimistic estimate.
3. If the proposed nourishment is carried out, what is likely to happen in the next big northeaster? Mild hurricane? Severe hurricane?
4. What is the erosion rate of the shoreline here during the last 10 years? 20 years? What is the erosion rate since 1938 (the year of the first coastal aerial photography in South Carolina)? Since the mid-1800s (when the first accurate South Carolina shoreline maps were surveyed by the old U.S. Geodetic Survey)? What will the beach look like in 2 years, 5 years, and 10 years?
5. What will the proposed solution do to the beachfront along the entire island? Will the solution for one portion of an island create problems for another portion?
6. What will happen if an adjacent inlet migrates or closes? What will happen if the tidal delta offshore from the adjacent inlet changes its size and shape or if the channel moves?
7. If the proposed erosion solution is carried out, how will it affect the type and density of future beachfront development? Will additional controls on beachfront development be needed at the same time as the solution?
8. What will happen 20 years from now if the inlet nearby is dredged for navigation? If jetties are constructed or enlarged two inlets away? If seawalls and groins are built on nearby islands?
9. What is the 50–100-year environmental and economic prognosis for the proposed solution to erosion if predictions of an accelerating sea level rise are accurate?
10. What are the alternatives to the proposed solution to shoreline erosion? Should the threatened buildings be allowed to fall in? Should they be moved? Should tax money be used to move them?
11. What are the long-range environmental and economic costs of the various alternatives from the standpoint of the local property owners? The beach community? The entire island? The citizens of South Carolina and the rest of the country?

Communities need to get a firm, clear statement from the local Corps of Engineers district office regarding the projected cost and life span of the proposed replenished beach. District personnel should explain their goals and describe past beach success in simple language using simple, clearly labeled diagrams showing high- and low-tide lines, location of the dry beach (above the high-tide line), and the storm berm. What is the plan if the beach disappears more quickly than anticipated? Who will pay if this happens?

Beach Scraping: You Can't Get Something for Nothing

Redistribution of beach sand by scraping has been used on several South Carolina beaches to provide storm protection, with varying amounts of success. As a means of coping with chronic shoreline erosion at Folly Beach and at Palmetto Dunes on Hilton Head Island, the

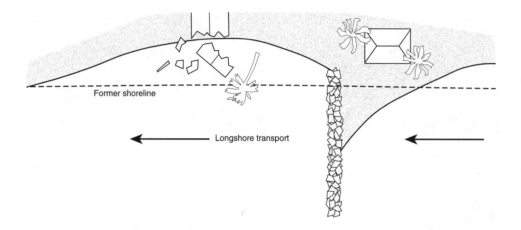

Former shoreline

Longshore transport

4.4. Groins trap sand on their updrift side, causing accelerated erosion and property loss in the downdrift area.

practice demonstrated diminishing returns to the environment. As a means of establishing a "seed" dune for Hurricane Hugo–ravaged beaches, the practice was largely successful on nonerosional shorelines. John T. Wells and Jesse McNinch reported how a scraped beach on Topsail Island, North Carolina, responded to Hurricane Hugo (appendix D, ref. 33).

Scraping the beach adds no new sand; in fact, "manicuring" might be a better name for it. Moving sand from the low-tide beach to the upper beach smoothes and widens the high-tide dry beach—an illusion in terms of added protection. Scraping is viewed as successful if the beach is smooth. However, the reality is that every time sand is manipulated, some is lost, never to return. Beach scraping is no longer permitted in South Carolina except in emergency situations.

In summary, beach replenishment upsets the natural system and is costly and temporary, requiring subsequent replenishment projects to remain effective. The Corps of Engineers refers to its beach replenishment projects as "ongoing," but "eternal" would be a better word. Also, serious economic questions can be raised when the facts associated with beach nourishment are considered, especially since it is not the general public—who pays for these projects through taxes—that typically receives the greatest benefit from them, but a relatively few beachfront property owners. Cries for beach nourishment projects usually come from those with direct economic interests associated with beach use—that is, developers and owners of houses, hotels, beachwear and gift shops, and other commercial ventures in the community, especially those whose buildings

are in danger of falling into the sea. Beach nourishment paid for with tax money is a form of government subsidy of such ventures. It would be unusual to hear beach visitors clamor for such a project. If the beach disappeared, the public would simply go elsewhere, which is what the commercial interests are most concerned about to begin with.

It should also be kept in mind that when buildings are abandoned or removed, no fundamental damage is done to the beach itself. It simply moves to a new location (see chapter 2). Although many problems are associated with beach nourishment, it is a more acceptable approach than hard stabilization.

Groins and Jetties (Hard Stabilization)

Groins (including terminal groins and groin fields) and *jetties* are walls extending into the ocean from the shore, perpendicular to the shoreline. Jetties, which may be very long (sometimes miles long), are intended to keep sand from flowing into a ship channel. Jetties usually come in pairs. Groins are much smaller walls built along and perpendicular to the beach, usually away from channels and inlets, that are intended to trap sand flowing in the longshore (surf zone) current (fig. 4.4). Because of shoreline problems associated with

4.5. Left: Groin field on the southern end of Pawleys Island. Although this feature is a spit built by net north-to-south longshore sand transport, on this particular day the longshore drift was in the opposite direction, and sand was building out (trapped) on the south sides of the groins. Groins do little to reduce the extreme-hazard risk on spits and may add to the risk potential. During Hugo the ebb flow of Pawleys Island Creek cut a new inlet across the spit between one of the groin pairs. New inlet formation is likely to occur again.

4.6. Right: Terminal groin on Midway Inlet at the north end of Pawleys Island. Although the groin may stabilize the inlet's position under nonstorm conditions, it reduces the inlet's ability to enlarge during storm ebb flow, possibly contributing to back-island flooding.

groins (discussed below), they are seldom built individually. Series of groins are called groin fields. Groin fields are present on many South Carolina beaches; there are extensive ones on Pawleys Island (fig. 4.5), Folly Beach, and Edisto Island. Groins can be made of wood, stone, concrete, steel, or nylon bags filled with sand.

Terminal groins are common in areas where beaches and property are threatened by an inlet that is changing its position, like Midway Inlet at the north end of Pawleys Island (fig. 4.6), which separates the island from Litchfield Beach. The mouth of Midway Inlet actively migrates up and down this reach, being alternately several hundred feet up into Litchfield

spit or adjacent to Pawleys' terminal groin as it is today (fig. 4.6). In the mid-1980s this groin was reinforced because there was a real danger that it would be flanked by the inlet and thus become part of Litchfield. The inlet's movement can be counted on to continue. At the south end of the island is Pawleys Inlet, which separates Pawleys Island from Debordieu Island. The spit, along with the rest of the island, was stabilized by a groin field in the 1950s.

Condominiums on the northern end of Fripp Island are built behind a terminal groin and a revetted wall next to an inlet. Likewise, groins have been constructed at the old Coast Guard (Loran) Station on the north end of Folly Island.

Both groins and jetties are very successful sand traps. If a groin is working correctly, more sand will be piled up on one side of it than on the other. However, the beach may be eroded on the downdrift side (fig. 4.4). A groin field has the same effect, but on a larger scale. Jetties also trap sand and cause erosion downdrift. Nowhere on the South Carolina coast—perhaps in the world—is sand starvation of a beach by jetties more dramatically demonstrated than in the effect the Charleston jetties have had on Morris Island and Folly Island. This is discussed further in chapters 1 and 6.

Seawalls, Revetments, and Bulkheads (Hard Stabilization)

Seawalls, structures built back from and parallel to the shoreline, are designed to receive the full impact of the sea at least once during each tidal cycle. Seawalls are present in almost every highly developed coastal area and are fairly common along the South Carolina coast. Other common structures in South Carolina are *bulkheads* and *revetments.* Bulkheads are a type of seawall placed farther away from the shoreline, in front of the first dune—or what was the first dune. They are, technically, de-

signed to hold land in place and are meant to take the impact of storm waves only; however, they quickly become de facto seawalls. Revetments are stone facings placed on eroding scarps or on the seaward sides of bulkheads to protect them from storm waves (fig. 4.7). They are made of loose stacks of large stones. A wave breaking on a stone revetment will lose part of its volume in the spaces between the

rocks, which reduces the erosive effect of the backwash of the wave and the impact on the dune or bulkhead. Sometimes the generic term *seawall* is applied to all of these structures.

In terms of shoreline management, building a seawall, bulkhead, or revetment is a drastic measure because these structures protect upland property at the expense of the public beach. This is why they were outlawed by

4.7. Bulkhead and revetment on former Captain Sams Inlet, Seabrook Island. Artificial closure of the inlet eliminated the erosion problem for this particular property.

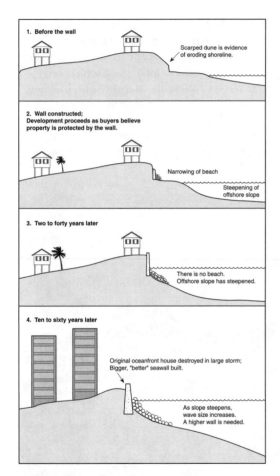

1. Before the wall

Scarped dune is evidence of eroding shoreline.

2. Wall constructed;
Development proceeds as buyers believe property is protected by the wall.

Narrowing of beach

Steepening of offshore slope

3. Two to forty years later

There is no beach.
Offshore slope has steepened.

4. Ten to sixty years later

Original oceanfront house destroyed in large storm;
Bigger, "better" seawall built.

As slope steepens,
wave size increases.
A higher wall is needed.

4.8. The saga of a seawall.

South Carolina in 1990 (see chapter 10). Seawalls harm the environment in four principal ways:

1. They reflect wave energy, ultimately removing the beach and steepening the off shore profile. The length of time required for this damage to occur is typically 1 to 30 or more years. The steepened offshore profile increases the storm-wave energy striking the shoreline, which in turn increases erosion.
2. They increase the intensity of longshore currents, hastening removal of the beach.
3. They prevent the exchange of sand between dunes and beach. Thus the beach cannot supply new sand to the dunes on the island, and the beach cannot flatten itself using sand from the dunes during storms.
4. They concentrate wave and current energy at the ends of the wall, increasing erosion at these points.

The emplacement of a seawall or other hard structure is usually an irreversible act with limited benefits. Since it gradually removes the adjacent beach, every seawall eventually must be replaced with a bigger, "better," more expensive wall (fig. 4.8). Although a seawall may extend the life of beachfront structures in normal weather, it cannot protect buildings on a low-lying barrier island from the damage caused by hurricanes. A seawall cannot prevent overwash

Table 4.2   Advantages and Disadvantages of Hard Stabilization (remember, seawalls are no longer a choice in South Carolina)

Advantages
  Temporarily protects property
  Low maintenance cost if properly constructed, especially in low wave energy settings

Disadvantages
  Eventually causes loss of the recreational beach (erosion due to wave reflection and refraction; cuts off sediment supply)
  Increases erosion at ends of wall and/or downdrift
  Limits access to beach
  Often ugly (loss of aesthetics)
  Can be costly (up to thousands of dollars per foot)
  Requires regular maintenance (additional cost)
  Debris from walls becomes hazardous on beach or to structures in back of wall during storms
  Creates a false sense of security, spurring additional development in high-hazard zone

or storm-surge flooding; in fact, floodwaters may be trapped and held behind the wall during a storm. And seawalls offer no protection against wind and windborne debris.

The long-range effect of seawalls can be seen in New Jersey and Miami Beach. In Monmouth Beach, New Jersey, in the late 1970s, the town building inspector told the au thors of the town's seawall history. Pointing to a seawall he said, "There were once houses and even farms in front of that wall. First we built small seawalls and they were destroyed by the storms that seemed to get bigger and bigger. Now we have come to this huge wall which we hope will hold." The wall he spoke of, adjacent to the highway, was high enough to prevent even a glimpse of the sea beyond. There was no beach in front of the wall; only remnants of old seawalls, groins, and bulkheads cluttered the once recreational shore.

A seawall is an expensive commitment designed to preserve shorefront property and structures only (table 4.2). The effect of seawalls and revetments on beach width and aesthetics in South Carolina can already be seen on parts of North Myrtle Beach, Myrtle Beach, Folly Beach, Seabrook Island, and Hilton Head Island (appendix D, ref. 91). We are fortunate that South Carolina's leaders took the bold step of outlawing these structures before all our beaches were destroyed.

Breakwaters (Another Kind of Hard Stabilization)

Old engineering designs die a slow death. As communities around the country have abandoned seawalls and groins in favor of conserving their beaches, they've sometimes turned to breakwaters, a poor alternative. Breakwaters interrupt and block the alongshore flow of sand, depriving downdrift beaches of their sand supply. Like groin fields and seawalls, breakwaters cause erosion to adjacent properties. Because they disrupt the wave pattern, unusual currents or wave patterns may form around them, usually leading to shore erosion focused at one or two sites. Drownings may be more frequent around breakwaters, and they present an identifiable situation where the community can be held liable for such accidents. Palm Beach, Florida, recently installed one of the new generation of offshore breakwaters known as a PEP reef. Within two years it had proved to be a failure, having caused increased shoreline retreat along much of the "protected" beach zone. At the time of this writing, the offshore breakwater is being removed. One final point: the price of maintenance for breakwaters can be enormous. The Beachfront Management Act, if adhered to, will prevent these structures along the South Carolina coast.

Other "Solutions": Selling Snake Oil on the Beach

Erosional crises tend to create a unique form of entrepreneurship. Shrewd businesspeople masquerading as coastal engineers or scientists promote all sorts of "solutions" to erosion problems. These so-called solutions range from artificial seaweed and various devices that block wave energy, suck water out of the beach, or trap sand, to dune cores built of cement structures or rubber tires lashed together.

Beware of promises that sound too good to be true, and of one device that claims to solve all erosion problems! When it comes to halting shoreline migration, only Mother Nature can lower the sea level. Shore-hardening structures and their low-budget cousins *do not* create sand.

Sea Level Rise: Built-in Obsolescence

If the techniques of shoreline stabilization have such a poor record, why are they used? Naturally, such structures were not intended to fail or to increase erosion. When the earliest ones were built, our understanding of shoreline processes was not what it is today. Thus, the impact the Charleston jetties would have on Morris Island and Folly Island probably was not anticipated; and in any case these islands were not cherished real estate in the 1800s. The older jetty systems were constructed *prior* to coastal development now perceived to be in need of protection. Furthermore, most engineering projects have a design life of less than 50 years; long-term geologic effects are simply not considered. In fact, a more common design life for shoreline-engineering projects is 10 to 20 years. The experiences of the Charleston jetties; the Cape May, New Jersey, jetties and seawalls; the Miami Beach groin field; and numerous other projects that represent long-term future commitments should tell engineers that they must consider the long-term consequences for all new projects. The consequences must be figured into cost-to-benefit ratios and should enter into the final decision of whether or not to pursue the stabilization project. The growing trend is against future hard stabilization projects, as is now the case in South Carolina, North Carolina, and Maine.

Several factors account for the long-range failure of shoreline stabilization schemes, but the most important and fundamental reason for failure is that the sea level is rising. In South Carolina this rise amounts to greater than 1 foot per century. All along the U.S. shore, what we are calling beach erosion is largely a response to the rising sea. As the sea level rises, barrier islands, in a long-term process (a very rapid process in geologic time, however), migrate landward. At the same time, of course, the mainland shore is eroding too; otherwise the islands would eventually bump into the mainland and would no longer be islands. (The Myrtle Beach area is an example of such welding onto the mainland shore by a barrier island. The attachment occurred thousands of years ago during an earlier high stand of sea level.) Island formation and migration depend on sea level rise, and under normal circumstances the rise does not threaten or endanger islands in any way.

What happens when we stabilize a barrier island beach? That is, what happens when we try to stop islands from migrating? Basically, shoreline engineering holds the beach, or the island, where it doesn't want to be held. The

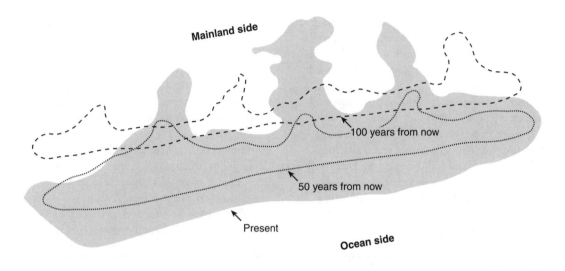

4.9. Hypothetical island responding to a rise in sea level.

beach or island is said to be out of equilibrium. It can also be said to be in trouble!

Figure 4.9 shows a hypothetical island responding to a hypothetical rise in sea level. The island becomes thinner and moves back. If the island is prevented from moving and is held in its original position, it will be in an increasingly precarious position as the years go by. As an extreme example, imagine that engineers had tried to hold an island in place when it first formed at the edge of the continental shelf 12,000 years ago. It would now be 200 to 300 feet under water, and the seawalls holding

back the ocean would have to be spectacular indeed! That is basically what coastal engineers are attempting to do today. They are trying to hold back a sea that is rising, but neither the older seawalls nor the newer beach nourishment designs take this phenomenon into account. Thus, obsolescence is built into these projects.

South Carolina is addressing this long-term crisis through the Beachfront Management Act, but the pressure to develop the shore is increasing. Like the original New Jersey resorts, many of our structures are not placed far enough back from the shore; nor have we been so prudent as to place all our structures behind dunes or on high ground. Consequently, our coastal development is no less vulnerable to the rising sea than New Jersey's, and as perceived problems grow, there will be political pressure to return to old habits. Hurricane Hugo provided a reminder. The solution lies in recognizing certain truths about the shoreline, not in trying to mold the shoreline to our concept of how it should be.

Truths of the Shoreline

South Carolina can learn from New Jersey and other coastal areas; certain generalizations, or "universal truths," about the shoreline emerge quite clearly from their experiences. These truths are equally evident to scientists who have studied the shoreline and old-timers who have lived there all their lives. As aids to safe, economic, and aesthetically pleasing shoreline development, these truths should be the fundamental basis of planning on any barrier island.

1. *There is no erosion problem until a structure is built on a shoreline.* Beach erosion (short-term shoreline retreat) is a common, expected event, not a natural disaster. Island migration is not a threat to barrier islands. Shoreline retreat is, in fact, an integral part of island evolution and the dynamic system of the entire barrier island. When a beach retreats, that does not mean that it is disappearing. Many undeveloped islands are migrating at surprisingly rapid rates, though only the few investigators who pore over aerial photographs are aware of it (examples are Waites Island, North and South Islands, and Capers Island). Whether the beach is growing or shrinking does not concern the visiting swimmer, surfer, hiker, or fisherman. It is only when we build a "permanent" structure in this zone of change that a problem develops.

2. *Construction on the shoreline causes the shoreline to change.* The location and shape of a sandy beach exist in a delicate balance with sand supply, wave energy, and sea level rise, a dynamic equilibrium. Most construction on or near the shoreline changes this balance and reduces the natural flexibility of the beach (fig. 4.7). The result is change, which threatens man-made structures. Dune removal, which often precedes construction, reduces the sand supply the beach needs to adjust its storm profile. Beach houses, even elevated houses, may obstruct the normal sand exchange between the dune-beach complex and offshore areas during storms. Similarly, engineering devices interrupt or modify the natural cycle (figs. 4.1–4.8).

3. *Shoreline engineering protects the interests of a very few, often at a very high cost in federal and state dollars.* Shore stabilization projects are in the interest of a very few beach property owners, not the public. If the shoreline were allowed to migrate naturally over and past the seawalls and beach houses, the fisherman and swimmer would not suffer. Yet beach property owners apply pressure to spend tax money—public funds—to "protect" the beach. Because these property owners do not constitute the general public, their personal interests do not warrant the large expenditures of public money required for shoreline stabilization.

Exceptions to this rule are beaches near large metropolitan areas. The combination of extensive high-rise development and heavy beach use affords ample economic justification for extensive and continuous beach replenishment projects. For example, it is far more reasonable to spend tax money to replenish the Grand Strand beaches, which accommodate tens of thousands of people daily during the summer months, than to spend tax dollars to replenish a beach that serves only a few private houses. In the case of the former, the beach maintenance is in the interest of the public that pays for it; in the latter case the expenditure amounts to upper-middle-class welfare. As coastal populations increase, however, and if the same mistakes are made at other locations, the frequency of cases requiring large expenditures will increase as well, as will the taxpayer's burden to underwrite these "eternal" projects. The shorter life, increased cost, and diminishing sand supply for each successive beach renourishment illustrate this truth.

4. *Shoreline engineering destroys the beach it was intended to save.* If this sounds incredible to you, drive to New Jersey and examine its shores. See the miles of "well-protected"

shoreline—without beaches (fig. 1.5)! If the truth be told, however, you don't really need to leave the state to see limited examples of a "New Jersey beach."

5. *The cost of saving beach property through shoreline engineering is usually greater than the value of the property to be saved.* Price estimates often prove to be unrealistically low for a variety of reasons. Maintenance, repairs, and replacement costs are typically grossly underestimated or even omitted, because it is erroneously assumed that the big storm, capable of removing an entire beach replenishment project overnight, will somehow bypass the area. The inevitable hurricane or northeaster, moreover, is viewed as a catastrophic act of God or a sudden stroke of bad luck for which one cannot plan. The increased potential for damage resulting from shoreline engineering is also ignored in most cost evaluations. In fact, very few shoreline-engineering projects would be funded at all if those controlling the purse strings realized that such "lines of defense" must be perpetual.

6. *Once you begin shoreline engineering, you can't stop it!* This statement, made by a city manager of a Long Island, New York, community, is confirmed by shoreline history throughout the world. Because of the long-range damage caused to the beach it "protects," shoreline engineering must be maintained indefinitely. When the sandy shoreline is not allowed to migrate naturally, the results are a steepening of the beach profile, reduced

sand supply, and therefore accelerated erosion. Thus, once a shoreline structure is installed, "better"—larger and more expensive—structures must subsequently be installed or more costly beach renourishment projects must be undertaken. In the end, however, all will suffer the same fate (fig. 4.8).

History shows us that there are two situations that may terminate shoreline engineering. First, a civilization may fail and no longer build and repair its structures. This was the case with the Vikings and the Romans; both built mighty seawalls. Second, a large storm may destroy a shoreline stabilization system so thoroughly that people decide to stop trying. In the United States, however, such a storm is usually regarded as an engineering challenge and thus results in continued shoreline stabilization attempts. After a destructive storm, we want the beach to be the way it was before the storm. This is the popular desire, irrespective of how many times the beach was engineered in the past or at what increasing cost. Also not considered in our short-term view is how soon it will need to be done again, how much bigger it will need to be, and what will be the cumulative cost.

Alternative Solutions to Coastal Engineering

There are alternatives to costly, and ultimately futile, shoreline engineering projects, but accepting them requires a different way of thinking about the coast.

1. Accept storms as a way of life on the coast. Property damage or loss is not caused by an "unexpected" or "unusual" storm; storms are the rule.
2. Design to live with the flexible island environment. Encourage dune building and conservation of vegetation, but don't fight nature with a "line of defense."
3. Consider all structures built near the shoreline temporary, even storm-resistant lighthouses.
4. Accept as a last resort any engineering scheme for beach restoration or preservation, and then only for metropolitan areas or as a temporary stopgap until a relocation plan can be implemented.
5. Base decisions affecting island development on the welfare of the public rather than that of a few shorefront property owners.
6. Let the lighthouse, beachfront home, or hotel be abandoned or moved when the time comes. Adopt a plan of systematic retreat (relocation).

The Relocation Alternative

When a building is threatened by erosion, the costs and benefits of moving it back from the shore must be weighed along with other alternatives (table 4.3). Depending on the nature of the problem, relocation can compare favorably with other alternatives and may prove to be economically and aesthetically better in the long run. For example, moving the Grand Strand's oceanfront hotels out of harm's way may not be economically or geographically feasible, but relocation is a viable alternative in many areas of the South Carolina shore developed with single-family homes (fig. 4.10).

Remember that the cost of moving back, although high, is likely to be a one-time expense, whereas hard and soft stabilization approaches will be repetitive expenditures. When combined with the cost of ongoing maintenance, the latter "solutions" are generally more costly than relocation.

Table 4.3   The Advantages and Disadvantages of Building Relocation

Advantages:
Removes threat to building
Allows natural shoreline processes to continue
Preserves the beach
Good possibility of one-time-only cost
Maintains aesthetics of beachfront

Disadvantages:
High cost
The site must be deep enough to allow suitable moveback, or an alternative site must be purchased
Structure must be of a type or design/construction that allows it to be moved; e.g., a wood-frame house is easier to move than a cinder-block house on a poured concrete slab
There must be an infrastructure that allows/ supports moving buildings

4.10. House being moved at Cherry Grove, North Myrtle Beach, after Hurricane Hugo.

One of the lessons taught by Hurricane Hugo is that a barrier island responds to storms and other coastal processes as a single system. In order to design a way to live with the flexible nature of a barrier island, one must first evaluate the entire island in terms of its physical processes, including its response to human activities. The second step is to focus on the characteristics of the individual building site. Keep in mind that no barrier island is safe from hurricane winds, storm-surge flooding, wave erosion, overwash, or inlet formation and migration. On the other hand, low, bare, rapidly migrating Edingsville Beach is more likely to suffer from these hazards than the inner-island forested beach ridges of the seaward-growing portion of Kiawah Island. The latter will be subjected to coastal hazards, too, but these will be less frequent and probably less intense than those experienced by the former over the short term. An island's elevation and forest cover offer immediate clues to its stability. Other characteristics of the island add further information (e.g., inlet formation and migration, overwash, and erosion rates); in general, these indicate how the island has responded to previous storms.

Another important aspect in evaluating overall island stability is the island's response to past human activity, particularly construction. An island is not necessarily less risky merely because it has already been developed or because there are stabilization structures in place. More likely, the opposite is true. Shoreline stabilization structures are static (immobile) obstacles to the dynamic (mobile) processes that make up the island system, and they disrupt the balance of the system. As a result, the island's natural flexibility is lessened and it begins to readjust to the new conditions. In other words, structures sometimes create conditions favorable to their own destruction. The presence of groins, seawalls, or revetments on the beach indicate that the shoreline is subject to erosion—now compounded by the stabilization structures. Such a shoreline certainly is to be avoided. Removing vegetation for purposes of construction or to get a better view of the sea may increase the potential for storm damage or create a blowing sand nuisance. Roads to the beach built through the dune line may act as overwash passes. Dune removal invites storm disasters because the elevation has been lowered and the protective buffer has been lost. Hurricane Hugo proved the effectiveness of high, wide dunes for preventing or greatly reducing property damage.

The political infrastructure of your prospective island also may have a strong bearing on its overall safety. Unchecked growth or unenforced building and sand dune protection codes are examples of social conditions that may create threats to health or safety. Overloaded sewage treatment systems, inadequate or unsafe escape routes, loss of natural storm protection, structures that are not stormworthy, and vulnerable utilities are but a few examples of politically derived development problems. Developers and realtors often take a "head in the sand" attitude and are reluctant to face the reality that the beaches are eroding, that the island's dynamics are disrupted, and that the density of development exceeds carrying capacity with respect to island processes. Potential buyers aren't encouraged by words like *hazard, risk, mitigation,* and *cost.* These terms apply islandwide, not just to individual properties. Eroding beaches, removed dunes, disappearing forest cover, beach nourishment projects, and failing seawalls are community-wide problems that all individual property owners pay for, whether or not their property is directly involved.

Once you are satisfied with the island's natural stability, its response to human development activities, and the political setting, site selection is the next important step.

Selecting Your Site: Playing the Odds

Human nature is such that we are willing to gamble if the potential reward is worth the risk. In the case of a barrier island, the rewards you gain are the amenities of the seashore and other coastal environments. The risk is losing your property. Like smart gamblers who know the odds and try to reduce the house advantage, beach house buyers can and should identify the natural odds of coastal hazards and act accordingly.

Structures placed in the least dynamic zones (stable areas subject to less movement or

Table 5.1 Parameters for Evaluating Site-Specific Risks from Coastal Hazards. This list is designed specifically for unconsolidated, potentially erodible shores, such as bluffed shorelines and sandy systems (e.g., barrier islands, barrier beaches)

	Risk Level		
	High	Moderate	Low
Site elevation	< 3 m	3–6 m	> 6 m
Beach width, slope, and thickness	Narrow and flat; thin with mud, peat, or stumps exposed	Wide and flat, or narrow and steep	Wide with well-developed berm
Overwash	Overwash apron or terrace (frequent overwash)	Overwash fans (occasional overwash)	No overwash
Site position relative to inlet or river mouth	Very near	Within sight	Distant
Dune configuration	No dunes (see Overwash)	Low or discontinuous dunes	High, continuous, unbreached ridge, dune field
Coastal shape	Concave or embayed	Straight	Convex
Vegetation on the site	Little, toppled, or immature vegetation	Well-established shrubs and grasses, none toppled	Mature vegetation, forested, no evidence of erosion
Drainage	Poor	Moderate	Good
Area landward of site	Lagoon, marsh, or river	Floodplain or low-elevation terrace	Upland
Natural offshore protection	None, open water	Frequent bars offshore	Submerged reef, limited fetch
Offshore shelf	Wide and shallow	Moderate	Steep and narrow

change) are less likely to suffer damage. When you can identify areas, rates, and intensities of natural physical activity, you have a basis for choosing a specific homesite. Consider, for example, an inland river and the flat areas (called the "floodplain") next to it. Even casual observation reveals that rivers flood occasionally. If you watch for a longer period, you may find that the time and size of the floods follow a pattern. Perhaps the area adjacent to the river is flooded every spring, but the entire flood-plain is flooded only every 5 to 10 years on average. Once or twice in a lifetime the flood is devastating, covering an area greater than the floodplain (although that once or twice may be closely spaced in time). With this information, it is possible to predict the frequency and size of the floods expected in a given area. Individual floods are described on the basis on the frequency of a given flood level. For instance, if water has reached a certain level only twice during a 100-year period of record, a flood rising to that level is called a 1-in-50-year flood. Unfortunately, this terminology leaves the impression that a similar flood will not occur for another 50 years. Floods are spaced in time randomly; they may occur in successive years or even in the same year. A better way of thinking about such floods is in terms of probability. A 100-year flood has a 1 percent probability of occurring in any given year. You could have two such floods in successive years and it would not preclude the 1 percent probability of having another the third year. When you are flipping pennies, you don't expect to see three heads in a row when you start with the first flip, but after two in a row, the chance that the third flip will be heads is the same as on the first flip—50:50. And while you are not worrying about the next 100-year flood, you may experience the 70-year flood or the 1,000-year flood of which we know nothing because we have no weather records of sufficient length to even predict it. Such big events have occurred and will again.

The same is true for the coastal zone and barrier islands. But the flooding there does not come from rivers. Instead, the flood is storm surge, the rising of the sea level during a storm.

Of course, building a house in a place that is flooded once every year, or even once every 10 years, would be foolish. Given the choice, you might rather locate where the likelihood of flooding is 1 percent in any given year (the likelihood of flooding being nearly 100 percent in the lifetime of the building). Whether to locate in a flood-prone area at all should be determined by how important it is to choose that site and by the level of economic loss you are willing to sustain.

The frequency and elevation of storm-surge flooding in coastal areas is somewhat predictable (table 5.1), although no one can predict exactly when a storm that causes flooding to a certain level will occur. Thus, if you expect a 1-in-25-year storm-surge flood level of 8 feet for a particular stretch of coast—that is, if you

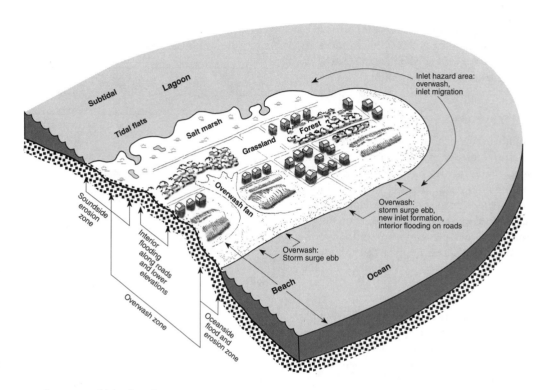

5.1. Summary of island environments, processes, and hazards to development. The development grid enhances these hazards by removing protective cover, creating avenues for overwash, and building at low elevations and too near the shoreline.

expect storm surge to reach a level of 8 feet once every 25 years—it would be sensible to build at an elevation greater than 8 feet. Because storm waves will further increase that height requirement, you should seek an even

higher elevation in addition to using a construction technique that raises the house several feet off the ground (see chapter 8).

The 100-year flood level is a standard used in both inland and coastal regions as the basis for the building codes and zoning ordinances adopted by communities that participate in the National Flood Insurance Program (see chapter 10). Flood level data and calculated wave heights are available from the Federal Emergency Management Agency (FEMA) or from community planning and insurance offices (Flood Insurance Rate Maps [FIRMS] show these elevations; see appendix B).

Stability Indicators: Some Clues for the Wise

A number of environmental attributes indicate the natural history of a given area. In revealing how dynamic an area has been through time, these attributes help prospective builders decide whether a site is safe for development. Such natural indicators include terrain (landform types), land elevation, vegetation type and cover, shells, and soil type (see table 5.1).

Terrain and Elevation

Terrain and elevation are good measures of an area's safety from various adverse natural processes. Low, flat areas are subject to destructive wave attack, overwash, storm-surge flooding, and blowing sand (fig. 5.1). The flooding of low areas is often from the landward side. Table 5.2 shows expected storm-surge levels for different parts of the South Carolina coast. Most areas less than 10 feet in elevation experience relatively frequent, severe flooding. Areas of 17-foot elevation received water damage from Hurricane Hazel in 1954. Hurricane Hugo, with its shore-perpendicular approach path and high rate of forward speed, created storm-surge levels higher than 20 feet in Bulls Bay. Hugo's surge completely covered

Table 5.2 Predicted and Measured Maximum Storm-Surge Levels for Several Points along the South Carolina Coast

Location	Predicted Storm-Surge by Hurricane Category (in feet)					Measured Storm-Surge Levels
	1	2	3	4	5	
Cherry Grove	5.8	9.8	13.9	14.3	18.0	Hazel 15 Oct 1954 (17.0) Hugo (13.0)
Myrtle Beach	5.8	10.6	14.4	15.2	19.2	Hazel 15 Oct 1954 (15.5) Hugo (13.9)
McClellanville	4.7	9.5	13.4	17.0	19.8	11 Aug 1940 (6.6) Hugo (16.4)
Isle of Palms	4.2	8.0	11.8	14.5	17.3	10 Aug 1940 (10.6) Gracie 29 Sept 1959 (7.9) Hugo (15.3)
Folly Beach	4.2	8.0	11.2	14.0	16.5	11 Aug 1940 (8.3) Gracie 29 Sept 1959 (8.7) Hugo (12.0)
Edisto Beach	4.7	8.3	11.5	14.4	17.2	27 Aug 1893 (11.2) 11 Aug 1940 (12.2) Gracie 29 Sept 1959 (11.9)
Hunting Island	4.6	7.9	11.1	13.8	16.6	11 Aug 1940 (8.0) Gracie 29 Sept 1959 (8.1)
Hilton Head	5.1	8.9	12.4	16.0	19.0	27 Aug 1893 (17.0)

Sources: Predicted levels and historic levels other than Hugo are from *Hurricane Evacuation Study for South Carolina*, pt. 2 (appendix D, ref. 19). Measured levels for Hugo are from *Storm-Tide Elevations Produced by Hurricane Hugo along the South Carolina Coast* (appendix D, ref. 22).

several islands, including Sullivans Island, Isle of Palms, and Folly Island (see chapter 3). Hugo might just as easily have come in 30 miles farther south and brought its most severe winds and highest storm surge to the densely populated Charleston metropolitan area.

Vegetation

Vegetation may indicate the environmental stability, age, and elevation of a site. In general, the taller and thicker the vegetation, the more stable the site and the safer the area for development. Maritime forests grow only at elevations high enough to preclude frequent overwash. In addition, since a mature mari-

time forest takes at least 100 years to develop, the homeowner is assured that such forest areas are generally the most stable and lowest-risk homesites on any island. The exceptions to this rule are areas where rapidly eroding shorelines have advanced into the maritime forest (e.g., at Debidue Beach, North Island, and Bull Island; see fig. 5.2) and where inlets are cutting into forested beach ridges—usually on the north ends of islands (e.g., Big Capers Island and Seabrook Island). A further advantage of building in a maritime forest is the shelter the trees provide from hurricane-force winds.

Typical maritime forests include live oak, long needle loblolly pine, red cedar, wax myrtle (the common shrub found in thickets),

5.2. Left: Eroding forest on the beach at Bull Island. The presence of maritime forest is not always a sign of stability. Photo courtesy of OCRM.

5.3. Right: The slope of the vegetation line reflects salt pruning of the maritime forest and thicket. This line of vegetation offers natural protection that should be retained. Unfortunately, builders often remove such vegetation during construction and replace it with nonnative species which neither stabilize the sediment nor offer wind protection.

cabbage palmetto, dogwood, black cherry, and holly. At the seaward edge the forest canopy will be lower and may be "pruned" by the salt spray (fig. 5.3).

Seashells

Seashells may also provide clues to the natural processes that have historically occurred in an area. A mixture of brown-stained and natural-colored shells often is washed onshore from the ocean side of islands during storms. Such shells thus indicate overwash zones. Don't build where overwash occurs; but if you must, elevate the building enough not to interfere with the process.

Mixed black and white shells without brown or natural-colored shells are almost certainly a sign that material has been dredged and pumped from nearby waterways (fig. 5.4). Such material is used to fill low areas on islands, to fill inlets that break through islands during storms, or to nourish eroding beaches. Thus, such a shell mixture may indicate an unstable area where development should be avoided. The oyster-shell-covered beach at Edisto Island is a good example of an old nourishment in an unstable area.

Soil Profiles

Soil profiles may give a clue to building-site stability. White-bleached sand overlying yellow sand to a depth of 2 to 3 feet suggests stability because such a soil profile (often found in forest areas) requires a long time to develop. Examine the soil profile by looking in a road cut, finger canal, or a pit that you have dug. Keep in mind that even formerly stable for-

ested areas can be eroded by a migrating barrier island, so you may find a "stable" soil profile in an unstable position. Avoid areas where profiles show layers of peat or other organic materials. Such layers have a high water content and lack the strength to support an overlying structure. The weight of a house can compress the layers, causing the house to sink. Furthermore, such soil conditions lead to septic tank problems.

Island Environments: Defining Levels of Activity

Inland developments typically occupy a single environment such as a pine forest or former pastureland. In contrast, barrier islands consist of small areas with very different environments, and typical developments overlap environmental boundaries without regard to the consequences. By knowing what environment(s) a building site or lot occupies you can identify the prevailing conditions, which may or may not be conducive to development. Typical barrier island environments include primary dunes, dune fields, overwash fans, grasslands, inlets, maritime forests and thickets, and marshes.

Primary Dunes

Primary dunes are usually defined as the row of dunes closest to the ocean, even though a distinct line or row may be absent. These

dunes serve as a sand reservoir that feeds the beach and back island, and provide a temporary line of defense against wind and waves.

The primary dunes are the natural main line of defense against erosion and storm damage to man-made structures, although it is a leaky line because of the overwash passes between the dunes. When development interferes with the dune system, both the natural and the built environments suffer. We must recognize the mobility of dune systems, even those stabilized by vegetation. Off-road vehicles, foot traffic, drought, and fire destroy vegetation, and therefore dune stability. In South Carolina, any activity that damages a primary dune is illegal. Conversely, the construction of wooden dune walkovers is encouraged—no state permit is required for them.

5.4. The color of shell staining may offer clues to a site's stability. Black shells are derived from eroded marsh muds; brown shells may indicate overwash.

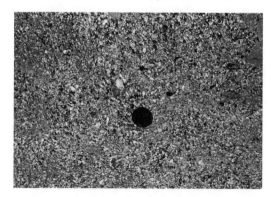

If dunes (oceanfront or interior) have been destroyed or threatened, there are some remedial steps you can take to stabilize them artificially. Planting dune grasses in bare areas helps stabilize the existing dunes and encourages additional dune growth (see appendix D, "Vegetation"). Sand fencing, elsewhere called snow fencing, is commonly used to trap sand and to increase dune growth.

The high elevation of nearby dunes does not in itself render a site safe. An area adjacent to the shore with a high erosion rate is likely to lose its dune protection during the average lifetime of a house. Even setback ordinances, which require that structures be placed a minimum distance behind the dunes, do not ensure long-term protection. In South Carolina, setbacks are based on historic erosion rates, but there is no guarantee that the past rate will be equal to future erosion rates. If you are located on a primary dune, you should expect to lose your home during the next major storm.

Dune Fields

Dune fields are open, bare to grassy dune areas. They are found between the primary dunes and the maritime forest (if present) or back side of the island. Stable dune fields offer sites that are relatively safe from the hazards of wave erosion, overwash, and storm-surge flooding, if the elevation is sufficiently high. However, digging up the dunes for construction may cause blowing sand and may destabi-

lize vegetation and increase sand movement.

We know that active coastal dune fields existed in South Carolina a century ago. Tuomey's 1848 *Report on the Geology of South Carolina* (appendix D, ref. 65) describes trees and houses being covered by dunes. One dune on the coast of Horry County "had already shut out from the house a magnificent view of the Atlantic, and its base had almost reached the yard gate."

Overwash Fans

Overwash fans (also called washover fans) develop when water thrown up by waves and storm surge flows between and around dunes. Such waters carry sand, which is deposited in flat, fan-shaped masses (figs. 2.3, 5.1). They also transport stained, bleached, and natural-colored shells to the inner island. Overwash fans provide sand to form and maintain dunes and build up the island's elevation. Where primary dunes are high and continuous, overwash is relatively unimportant and restricted to the beach and nearshore area. This is the case along much of the Horry County shore and on large islands such as Isle of Palms, Kiawah Island, and Hilton Head Island. Where dunes are absent, low, or discontinuous, overwash fans may extend across the entire island, as they do on Cape Island, Edingsville Beach, and Capers Island. During severe hurricanes, only the highest elevations (generally above 15–20 feet) are safe from

overwash. Hurricane Hugo caused extensive overwash along much of the northern half of the South Carolina coast. The overwash fans on low-elevation islands were so extensive that they coalesced in sheets called overwash terraces (fig. 5.5).

Overwash may damage or bury structures and roads (fig. 5.6) and block escape routes. Level roads that are cut straight to the beach often become overwash paths during storms, especially where the roads cut through dunes. Thus the roads built to expand development may contribute to its destruction. Such roads should end landward of the dune. Emergency vehicle accesses should go over the dunes. This concept was well illustrated at shore-perpendicular road ends during Hurricane Hugo on Folly Beach and Garden City (see chapter 3).

Avoid building on overwash fans, especially if they are fresh and unvegetated, although such areas may be difficult to recognize if the fans have been destroyed by bulldozing or sand removal. If no alternative site is available, elevate the building and allow the overwash to continue and build up sand. Use overwash deposits removed from roads and driveways to rebuild adjacent dunes; do not remove the sand from the area.

Grasslands

Grasslands are found behind dune fields and on the back side of islands, just inside the salt marsh. Such areas may be relatively flat, hav-

ing been built up as an overwash terrace, and generally are subject to future flooding and overwash. Natural grassland may be difficult to distinguish from artificial bulldozer-flattened developments, but the former is characterized by a diversity of plants (e.g., marsh grasses, yucca, cactus, and thistle).

Inlets

Inlets are the rivers or channels that separate islands. As a hurricane approaches a barrier island, strong onshore winds drive storm-surge waters and waves against the island and up the estuaries. As the storm passes, the wind either stops blowing or shifts to blow seaward, caus-

ing surge waters to return seaward. If the existing inlets do not allow the water to escape fast enough, a new inlet is cut from the marsh side.

Low, narrow island areas that lack extensive salt marsh and are near the mouths of rivers or estuaries are likely spots for inlet development. Spits, common on the southern ends of the sea islands and of Litchfield, Pawleys, and Folly Islands, are particularly vulnerable and should be avoided as building sites.

Once formed, an inlet tends to migrate laterally along the barrier island, although there is wide variation in individual inlet behavior through time (see appendix D, refs. 32, 59, 68, and 114, for directions and rates of inlet migration in South Carolina). Structures and

5.5. Left: Overwash terrace comprising coalesced overwash fans along the Cape Romain shoreline. The overwash fans are burying the salt marsh as the island migrates. Photograph by Rob Thieler.

5.6. Right: View looking south in Garden City after Hugo, showing extensive damage and road buried beneath overwash. Photograph by Rob Young.

property in the path of a migrating inlet are destroyed. Sediment carried seaward through the inlet builds an underwater ebb-tidal delta. In time the delta may fill the inlet, closing it naturally. Artificially filled inlets, such as the one that filled on Pawleys spit after Hugo, are not stable areas for development.

Maritime Forest, Thicket, and Shrub Areas

Maritime forest, thicket, and shrub areas are generally the least risky building sites. Under normal conditions, overwash, flooding, and blowing sand are not problems in these environments. The plants stabilize the underlying sediment and offer a protective screen. The problem is that building in these low-risk sites disturbs the vegetation, destroying the very aspect that helps reduce risk!

If you are building in a vegetated area, preserve as much of the vegetation as possible, including the undergrowth. Trees tend to buffer winds and are excellent protection from flying debris during hurricanes. Remove any large dead trees from the construction site, but conserve the surrounding forest to protect your home (fig. 2.12). Stabilize bare sand areas as soon as possible with new plantings because bare sand is the first to erode by wind or water. The presence of an active dune field on the margin of a forest may threaten the stability of forest sites.

Marshes

Marshes (figs. 2.16, 5.1) are the breeding grounds for many fish and shellfish. Their extensive shallows provide considerable protection against wave erosion of the back side of the island. In the past, marshes were sometimes filled to provide more land on which to build, but this practice is no longer legal. Areas around finger canals were often built up with the material dredged from the marsh to form the canal.

Nature usually takes its revenge on those who occupy this land. Buried marsh provides poor support for building foundations and does not provide a quality groundwater reservoir. Thus, such building sites typically have an inadequate supply of fresh water, and their septic systems do not function properly. In addition, effluent waste from such sites has closed adjacent marshes to shellfishing. These are low-elevation areas susceptible to flooding from even minor storms.

By law, marshes can no longer be dredged or filled. It is often possible to plant *Spartina*, one of the marsh grasses, and create new marshes to stabilize areas that might otherwise turn to mudflats and lead to high-ground erosion. This method is highly preferable to bulkheading because it protects the shoreline while forming new habitat for marine plants and animals.

Water Problems: An Invisible Crisis

One of the most significant hazards to barrier island living is contaminated water. Basically, water contamination involves three factors: water supply, waste disposal, and any form of island alteration that affects the first two factors. Dredge-and-fill operations (e.g., inlet cutting, channeling islands for canals and water-ways, and piling up dredge spoil) and other human construction activities may alter the groundwater system.

Water Supply

Just as the quality and availability of water determine the plant and animal makeup of an island's ecosystem, they also determine, at least in part, the island's capacity to accommodate people. Water quality is measured by potability, freshness, clarity, odor, and the presence or absence of pathogens (disease-causing bacteria). Availability implies the presence of an adequate supply. Unless an island has a public water supply, however, this may be highly seasonally dependent.

In the absence of a municipal water system, the only fresh water directly available to a barrier island is the rain that falls on it. This rainwater seeps through the porous and permeable sands and builds up as a lens or wedge of fresh water. This lens overlies salt water, which seeps into the sediments from the adjacent ocean, inlet, or marsh. The higher the island's elevation above sea level and the greater the accumulation of fresh water, the greater the thickness of the freshwater lens. If clean sands underlie the island, the thickness of the lens should be about 40 feet for every 1 foot of average island elevation. This is rarely the case in South Carolina, however, because dense clay underlies many of the islands. The top of the freshwater lens is known as the groundwater table, and on many islands this shallow reservoir is the only supply of domestic fresh water.

If too many wells are dug into the groundwater table, the table drops. Early occupants of a development should not be surprised when their shallow wells dry up as the development grows. If many wells are overpumped and the groundwater table goes down, salt water intrusion may occur. Seeking alternative sources of water such as deep aquifers or building alternative sources such as municipal water systems (e.g., deep wells, pipelines, filtration plants, and desalinization plants) is expensive and is done only on densely developed islands.

Large developments draw their water supply from rock units beneath the younger surface sands and muds. These aquifers are formed by rock formations that are exposed on the coastal plain (their recharge area) and dip seaward beneath the coast. The fresh water in such aquifers has been accumulating over thousands of years, but large developments withdraw it faster than it can be replaced (recharged). In effect, the water is being mined, and as the fresh water is pulled out, the space is refilled with salt water, contaminating existing wells and destroying the adjacent aquifer as a source of fresh water. This occurred early in the 1900s at the U.S. Marine Corps training center on Parris Island.

Hilton Head Island exemplifies high water use on the South Carolina coast. This island uses close to 40 percent of all the groundwater pumped in the South Carolina low country, an average of 10 million gallons a day! Savannah, Georgia, pumps 45 million gallons a day from the same aquifer. As a result, Atlantic Ocean water is moving into the principal artesian aquifer, and much of the area under Hilton Head Island now shows saltwater concentration. The salt content is increasing in the aquifer beneath Parris Island, Fripp Island, Edisto Beach, and the more inland parts of Beaufort, Colleton, Jasper, and Hampton Counties (appendix D, refs. 123, 124). Large communities may address the problem of expensive water treatment with solutions that are not possible for smaller islands.

As condominiums and high-rises replace cottages, the demand for water will increase. Alternative sources must be sought, and they may be expensive. Island property owners will bear the cost. Consult the proper authorities about water quantity and quality before you buy (see appendix B, "Water Resources")!

Waste Disposal

Wastewater disposal goes hand in glove with water supply. On many South Carolina barrier islands the home septic system is the primary means of wastewater disposal. This system consists of a holding tank in which solids settle and sewage is biologically broken down, and a drain field that allows water to percolate into the soil. The soil then filters and purifies the water. Sometimes the same natural system that is used to cleanse the water is also used to supply potable water to residences.

Many communities are unaware of the potential water problems they face. Crowded development, improperly maintained septic systems, and systems installed in soils unsuitable for filtration have resulted in poorly treated or untreated sewage entering the surrounding environment. Polluted water is a primary source of hepatitis and other diseases. Sewage may also enter sounds and marshes, contaminating shellfish and ultimately destroying that resource. It is a sad irony that we have the understanding and technology to make environmentally safe septic systems, but we sometimes fall short of the enforcement capability to ensure that they are always used.

Public officials should require strict enforcement of existing codes, strict policing of existing systems, and proper site evaluation before permits are issued. Homeowners should learn the mechanics of septic systems to prevent malfunctioning or to spot it early (see appendix B, "Sanitation and Septic System Permits"). Municipal waste-treatment plants may be an answer for the larger communities, although such plants may become overloaded or inefficient. Contact your local health department for more information.

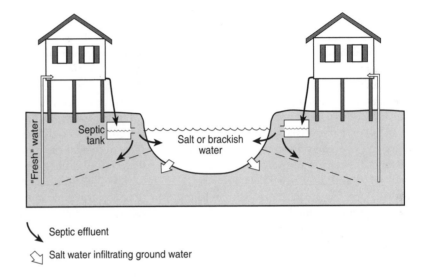

Septic effluent

Salt water infiltrating ground water

Finger Canals

A common island alteration of the 1960s and 1970s still causing island water problems are finger canals (figs. 5.7, 5.8), channels dug into an island for the purpose of providing additional waterfront lots. Examples of finger canals can be found at Cherry Grove, Garden City Spit, and Harbor Island.

Six major problems are associated with finger canals: (1) the lowering of the groundwater table, (2) pollution of groundwater by seepage of salt or brackish canal water into the groundwater table, (3) pollution of canal water by septic seepage into the canal, (4) pollu-

tion of canal water by stagnation resulting from lack of tidal flushing or poor circulation with sound waters, (5) fish kills generated by higher canal water temperatures, and (6) fish kills generated by nutrient overloading and deoxygenation of the water. Compounding these problems, finger canals in South Carolina quickly fill with sediment and become unusable or unnavigable. This leads to requests for dredging, which is expensive and may not be permitted for the above reasons.

Bad odors, flotsam of dead fish and algal scum, and contamination of adjacent shellfishing grounds are symptomatic of pol-

5.7. Left: Finger canal, northern North Myrtle Beach.

5.8. Right: The finger canal saga.

luted canal water. Thus, finger canals often become health hazards or simply too unpleasant to live near. Residents along some older Florida finger canals have built walls to separate their homes from the canals.

If you consider buying a lot on a canal, remember that canals usually are not harmful until houses are built along them. Short canals, a few tens of yards long, are generally much safer than long ones. Also, although most canals are initially dug deep enough for small-craft traffic, sand movement on the back sides of barrier islands fills in the canals, causing subsequent navigation problems. Most of the finger canals in South Carolina are sediment choked and are virtually unnavigable to anything larger than a johnboat.

Finally, on narrow islands, finger canals dug almost to the ocean side offer a path of least resistance to storm waters and are therefore potential locations for new inlets. Property owners along finger canals on Dauphin Island, Alabama, found themselves owning tiny islands or open water after Hurricane Frederic came through in 1979. Because of the multitude of problems associated with finger canals, they are now rarely, if ever, permitted in South Carolina.

Site Safety: Rules for Survival

In order to determine the safety of a site on a barrier island, it is necessary to evaluate all of the prevalent dynamic processes on the island. Information on storm surge, overwash, erosion rates, inlet migration, longshore drift, and other processes can be obtained from maps, aerial photographs, scientific literature, and personal observation. Appendix D provides an annotated list of scientific sources, many available at little or no cost; you are encouraged to obtain those of interest to you. Although developers and planners usually have the re-

sources and expertise to use such information in making decisions, they sometimes ignore it. In the past, the individual buyer was not likely to seek the information necessary to decide on the suitability of a given site. Today's buyer should be better informed.

Buyers, builders, and planners can assess the level of risk they are willing to take with respect to coastal hazards. The list of specific dangers and cautions below provides a basis for taking appropriate precautions in site selection, construction, and evacuation plans. Our recommendation is to avoid extreme- and high-risk zones. Keep in mind, however, that small maps of large areas are generalized, and that every site must still be evaluated individually. Safe sites may exist in high-risk zones, and dangerous sites may exist in low-risk zones.

The following characteristics are essential to site safety:

1. The site elevation is above the anticipated storm-surge level (table 5.1).
2. The site is behind a natural protective barrier such as a line of sand dunes.
3. The site is well away from a migrating inlet.
4. The site is in an area of shoreline growth (accretion) or low shoreline erosion. Evidence of an eroding shoreline includes (a) a sand bluff or dune scarp at the back of the beach; (b) stumps or peat exposed on the beach; (c) slumped features such as trees, dunes, or man-made structures; and (d) protective devices such as seawalls, groins, or replenished sand.
5. The site is located on a portion of the island backed by healthy salt marsh.
6. The site is away from low-elevation, narrow portions of the island.
7. The site is in an area of no or infrequent historic overwash.
8. The site is in a vegetated area that suggests stability.
9. The site drains water readily.
10. The fresh groundwater supply is adequate and uncontaminated. There is proper spacing between water wells and septic systems.
11. The soil type and elevation are suitable for efficient septic tank operation.
12. No compactable layers such as peat are present in the soil (the site is not a buried salt marsh).
13. Adjacent structures are adequately spaced and of sound construction.

Escape Routes

The effective and early evacuation that preceded Hurricane Hugo set a good example for future storms. Most of the population of Charleston, Georgetown, and Horry Counties relocated inland; as a result, few lives were lost. The traffic was bumper to bumper on the few roads leading west. (Accidents may back up traffic for miles.) Motel lobbies were filled with people looking for a place to stay, but all the rooms were taken. Weary people were forced to continue traveling until they found space. Similar experiences should be anticipated for the inevitable future hurricane in South Carolina, for coastal populations have grown considerably. If you have driven in the Charleston area during normal rush hour, try to imagine this low-lying city, its suburbs, and the adjacent islands evacuating in the face of a hurricane! Don't wait. Leave early.

There should be a route that will permit escape from an island to a safe location inland within a reasonable period. The presence of a ready escape route near a building site is essential to site safety, especially in areas with high-rises, where the number of people to be evacuated, transported, and housed elsewhere is large.

Select an Escape Route Ahead of Time

Check to see if any part of your potential escape route is in a low area subject to blockage by overwash or flooding; if so, seek an alternate route. Several exit routes from South Carolina islands are flood-prone. Note whether there are bridges along the route. Remember that some residents will be evacuating pleasure boats and that fishing boats will be seeking safer waters; thus, drawbridges will be accommodating both boats and automobiles. Periodically reevaluate the escape route you have chosen—especially if the area in which you live has become more congested. With more people using the route, it may not be as satisfactory as you once thought.

Use the Escape Route Early

Be aware that several South Carolina islands have only one route for escape to the mainland, and that in some cases exit is via another island (e.g., Fripp Island via Hunting Island). In the event of a hurricane warning, leave the island immediately; do not wait until the route is blocked or flooded. Anyone who has experienced the evacuation of a community knows of the chaos at such bottlenecks. Depend on it: excited drivers will cause wrecks, run out of gas, and have flat tires; and cars of frightened occupants will be lined up for miles behind them. Be sure to have plans for where you will go. Keep alternative destinations in mind in case you find your original refuge filled or in danger. Finally, find a place of last refuge where you can go rather than being stuck in your car should the situation arise. Parking garages above the first floor and the lee side of high-rise condominiums and hotels are good bets. Remember to plan, and then follow your plan. Nothing less than your life is at stake.

As more and more people live and vacation along the coast, more lives, property, and dollars are put at risk. Our observations of barrier island communities after several hurricanes and winter storms suggest that property damage can be lessened significantly by prudent site selection and proper location of structures. Until recently, however, there was no established process to effectively and routinely evaluate the potential risks of storm damage to development. A new approach to coastal hazard assessment considers the risk of damage to coastal development from hurricanes and winter storms based largely on a site's geologic setting (appendix D, ref. 92).

This new technique, known as *coastal risk mapping*, uses coastal physical processes and island geomorphic characteristics to rate the overall risk of storm damage at a given site as "low," "moderate," "high," or "extreme" (table 6.1). A preliminary evaluation maps the island's elevation zones and forest cover. Next, additional criteria are considered: the island's width; its interior sand dune height, width, and distribution; its potential for inlet formation; modern inlet dynamics; its historic storm response; the engineering structures and projects present; and other human modifications of the natural environment. The preliminary evaluation is adjusted to reflect these supplementary criteria, and a final risk map is produced that reflects the relative risk for property damage as outlined below.

Extreme-Risk Zones

Low elevation, within the 100-year flood level, and exposed to open ocean or a wide lagoon so that waves greater than 3 feet are likely (equivalent to FIRM V-zones). Vegetation consists of sparse growths of low beach or dune grasses. Extreme-risk areas are likely to be flooded by storm surge waters and heavy rains, and are likely to be overwashed during storms. These areas are the most susceptible to wave attack and storm-surge ebb scour, and often are affected by erosional shoreline retreat. Extreme-risk areas are commonly, but not exclusively, oceanfront.

High-Risk Zones

Low-elevation areas within the 100-year flood zone (FIRM A-zones) susceptible to flooding and wave attack. Vegetation is usually sparse, similar to that found in extreme risk zones. High-risk areas are likely to be flooded by storm-surge waters and heavy rains; overwash is less likely, though possible. These areas are only slightly less susceptible to wind damage than extreme-risk zones. They are usually located farther inland, however, and thus are usually less susceptible to erosion damage.

Moderate-Risk Zones

Above the 100-year flood zone (FIRM B-zones or higher) but lacking maritime forest or dense shrub thicket cover; not generally subject to flooding and unlikely to suffer direct wave attack. Areas that lie within the 100-year flood zone (FIRM A-zones) but have a vegetation cover such as dense maritime forest growth or shrub thicket that significantly reduces the impact of wind are considered moderate-risk zones.

Low-Risk Zones

Above the 100-year flood zone (FIRM B-zone or higher) and well forested, generally not subject to flooding and wind hazards. Removing forest for development will reduce the amount of protection and lead to increased degradation of newly exposed portions of the forest from salt spray.

Virtually all coastal areas are at extreme risk for a category 5 hurricane! Differentiation into risk zones is useless for such a storm. The relative risk zones presented here are based on the risk afforded to property by a low category 3 hurricane hitting directly at the site in question. A low category 3 hurricane will have winds of about 111 to 120 miles per hour and will be accompanied by a storm surge ranging from 5 to 12 feet, depending on coastal configuration and offshore bathymetry.

Risk maps facilitate the assessment process and help determine viable property damage mitigation alternatives. Ultimately, mitigation measures will help reduce dollar losses, resulting in more affordable insurance for the property owner and easing the burden on government disaster response and recovery programs.

Note that coastal risk mapping considers only the risk of property damage; risk maps do

Table 6.1 Characteristics of Coastal Property Damage Risk Categories. Also included are brief descriptions of the type of damage that might be expected from a storm and recommendations for mitigation of property damage

RISK	Primary Determinants of Vulnerability (determine storm surge, flood, overwash, and wind hazards)		Secondary Determinants of Vulnerability			Vulnerability	Probable Damage Destruction in Category 3 Hurricane of Equivalent Wind/Wave/Surge	Mitigation
	Elevation	Vegetation	Erosion Rate	Inlet Potential	Construction Factor			
EXTREME	Low; mostly in V zone; some A zone sites. Typically back beach, low frontal dune, active overwash front, similar on sound side and near inlets	None or sparse, beach/dune grasses only; no maritime forest or shrub thicket	High to moderate	Near migrating, historic, or potential inlet position	Older buildings built before building codes, not to code, or with code violations, closely spaced, or new buildings that were not built to code or violate code	MAXIMUM (likely to be impacted by 4 or more hazardous processes)	Total destruction to very heavy damage by direct wave attack for all elements of buildings (roof, windows, walls, foundations, decks, porches, stairs, garages) as well as outbuildings, utilities, services, and landscaping	Relocate before storm or abandon site after storm. *Do not rebuild destroyed buildings in place.* Protect marshes. Preserve, augment, and restore all natural environments
HIGH	Low; in A zone and lacks forest or shrub thicket cover. Typically overwash apron or fan extension but away from V zone; inner dune trough or frontal blowout in dunes, perhaps flat interior grasslands	Sparse to none, or greatly disturbed by development or past salt-water kills	High to moderate	As above	As above	HIGH (likely to be impacted by at least 3 hazards)	Total destruction is not uncommon and heavy damage is likely; wind damage and flood damage most probable; all building elements at risk (roof, windows, walls, foundations, attachments) as well as outbuildings, services, and landscaping	Relocation is most prudent; elevate and vegetate; build protection in region around site (e.g., dune projects, forestation), plug dune gaps, change street layout; maintain existing projects (e.g., beach nourishment), but not shore hardening. Protect marshes

Table 6.1 (continued)

RISK	Primary Determinants of Vulnerability (determine storm surge, flood, overwash, and wind hazards)		Secondary Determinants of Vulnerability			Vulnerability	Probable Damage Destruction in Category 3 Hurricane of Equivalent Wind/Wave/Surge	Mitigation
	Elevation	Vegetation	Erosion Rate	Inlet Potential	Construction Factor			
MODERATE	Moderate to high, and *not* in V zone; possibly A zones if heavily forested. B zone or out of flood zone (but downgraded by secondary hazards). In dunes or other elevated landforms such as built-up terrace	Stable vegetation of maritime forest or shrub thicket (not seriously disturbed by development), interior and frontal dunes well-vegetated, island backed by healthy marsh. A zones even with good vegetative cover can still be at high risk	Low to accretionary	Should be away from migrating inlet; downgrade if near historic or potential new inlet	Downgrade if buildings at high risk. All buildings should meet or exceed building code standards	MODERATE (likely to be impacted by at least 2 hazards, and nuisance hazards such as overwash and dune migration are likely)	Heavy to moderate damage should be expected. Wind damage very likely (e.g., windows, roof, walls, attachments). Flood damage less likely. Landscaping affected by overwash and blowing sand. Above ground services likely to be interrupted	Elevate and vegetate onsite surrounding region (e.g., vegetate dunes, rehabilitate forest, plug overwash gaps or work to maintain natural overwash sites); promote sand additions to island interior; modify street layout. Protect marshes
LOW	High, C zone (or B zone but upgraded by healthy maritime forest and no secondary problems)	Maritime forest cover is healthy and largely undisturbed. Surrounding environs also well-vegetated	Zero to accretionary	No inlet potential	All construction is at or above code and is well-maintained	LOW (no more than one hazard likely)	Expect damage, but not heavy. Wind damage is most likely (cosmetic to exterior roof, walls, windows, attachments). Potentially serious damage from falling trees or blowing debris. Rain damage if glazing fails or roof/wall leaks develop	Augment protective vegetation and dunes as above. Do not remove sand. Protect marshes

Modified from *Living by the Rules of the Sea* (appendix D, ref. 92).

not predict risks to inhabitants. In general, of course, areas with a high potential for property damage are also areas of high risk for human inhabitants, but some risks have nothing to do with homesite safety—lack of a good evacuation route, for example. Inhabitants trapped or isolated by a great storm are at extreme risk regardless of the vulnerability of their property.

A great deal of information is condensed into the coastal risk maps (RMs), including data from the Office of Ocean and Coastal Resource Management's (OCRM) orthophoto maps transferred to digital base maps created by the U.S. Census Bureau. Coastal setback information is not visible at the scale of these maps and thus must be obtained from the original sources; but the shoreline erosion zones are accurately portrayed on each risk map.

Three types of erosion zones are shown on the risk maps: standard erosion zone (S), unstabilized inlet (Iu), and stabilized inlet (Is). These designations indicate how the baseline was determined for that area and what physical forces control the area's erosional characteristics. For further information, refer to the discussion of the Beachfront Management Act in chapter 10 and appendix C.

Now that you have the tools with which to assess the historical and active processes, coastal hazards, vulnerability, and risk of a coastal area, you can evaluate your own area of interest, be it a favorite island, community, neighborhood, existing house, or building site for your island dream home. The following island-by-island descriptions and accompanying risk maps (RMs) of the South Carolina coast include storm impact history, patterns of shoreline engineering, local anecdotes, definition of hazardous areas, identification of areas likely to develop major problems in the future, and, where appropriate, some information on the impacts of Hurricane Hugo. Each description can be used as a guidebook to examples of the good, the bad, and the risky of island development. Risk maps are produced in a large-scale format, so some detail is lost when they are produced at the smaller scale of this book. Full-size maps will be available in an atlas of hazard risk maps planned for the near future.

The Grand Strand

The northernmost segment of South Carolina's coast is an arcuate strand commonly known as the Grand Strand. Lying between Cape Fear, North Carolina, and Cape Romain, South Carolina, this coastal segment is one of three great crescent shorelines that exist between the capes, extending to the north. Crescent shorelines are typical of microtidal coasts whose shore characteristics are determined by wind-driven waves. The arcs result because longshore currents alternately drive sediment upcoast in the warmer months and downcoast during the winter.

Because the tidal range is low (less than 6 feet) the barrier islands are relatively long and narrow, and the inlets that separate them are generally small. The only exceptions are found at the mouths of large rivers.

Horry County (RMs 1, 2, 3, 4, 5, and 6)

This northernmost coast of South Carolina extends from the North Carolina–South Carolina border in Little River Inlet to just south of the community of Garden City. With the exception of Waites Island, approximately 50 miles of continuous beach mark the edge of the mainland coast known as the Grand Strand. Immediate beachfront property is at risk, but the flood zone is relatively narrow (often less than 0.5 mile) compared with the rest of the South Carolina coast. Erosion rates are generally low to moderate (less than 3 feet/year); however, the developed communities have held their shorelines in place through either older hard stabilization or beach nourishment. South of Hog Inlet there are a few swashes (natural drainage lines through low areas), but the zone of overwash and storm-surge wave impact is small. Storm evacuation should not be a problem here, although you should not allow this to generate a false sense of security. Evacuation is still a necessity in the face of a storm, and population growth means that more time is needed for such emergency procedures. Beachfront property is at high risk, and erosion is a growing problem. The risk maps that follow proceed from north to south.

Waites Island (RM 1)

Waites Island (also shown on maps as Waties, Waiter, and Wader Island) is a 3-mile-long island owned in part by the University of South Carolina, Coastal Carolina College, in Conway. Currently undeveloped, the island is a good example of an extreme-risk area that certainly is being eyed for potential development. Its proximity to North Myrtle Beach, good water access, unspoiled beach, and shrub pine forest cover would look good on a realty brochure. But there are serious problems here. The small size of the island and the fact that it is bounded by inlets mean that it is highly unstable. The jetty stabilization of Little River Inlet and continued engineering of the south side of Hog Inlet may adversely affect the natural sand supply for the island. The erosion rate is very high (3–9 feet/year). Evidence of beach erosion is common in the form of beach scarps, slumped shrubs, and fallen trees at the back of the beach; however, as long as there is no development this is not a problem. Hurricane Hugo overwashed part of the island. There are active dunes on the island, and some dunes have become reactivated since their vegetation cover was lost to forest fire.

If development does take place, it should be on a very limited scale and restricted to the highest elevations near the middle of the is-

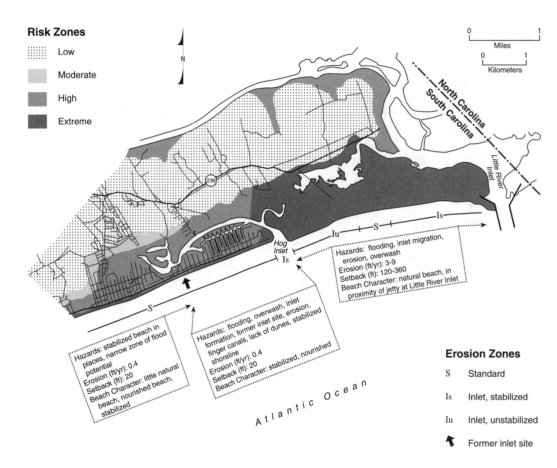

Risk Zones

Low

Moderate

High

Extreme

Hazards: stabilized beach in places, narrow zone of flood potential
Erosion (ft/yr): 0.4
Setback (ft): 20
Beach Character: little natural beach, nourished beach, stabilized

Hazards: flooding, overwash, inlet formation, former inlet site, erosion, finger canals, lack of dunes, stabilized shoreline
Erosion (ft/yr): 0.4
Setback (ft): 20
Beach Character: stabilized, nourished

Hazards: flooding, inlet migration, erosion, overwash
Erosion (ft/yr): 3-9
Setback (ft): 120-360
Beach Character: natural beach, in proximity of jetty at Little River Inlet

Erosion Zones

S   Standard

Is   Inlet, stabilized

Iu   Inlet, unstabilized

➤   Former inlet site

RM 1. Waites Island, Cherry Grove, and northern North Myrtle Beach

6.1. Condominiums on Hog Inlet at the north end of Cherry Grove are at extreme risk. Left: The small stones that form the revetment may become missiles hurled against the buildings by waves of future storms. Storm flooding, wave erosion, and inlet migration are all possible. Right: The density of development exceeds the carrying capacity of the environment. Property loss in this area will be repeated in the future.

land, although such limited development might not recover the costs. Property damage is a certainty in a category 3 hurricane.

Northern Portion of Cherry Grove Beach (RM 1)

The northern portion of Cherry Grove Beach (Hog Island on older maps) from Hog Inlet (fig. 6.1) to south of the position of the old Cherry Grove Pier (about even with 60th Avenue North) may be regarded as a barrier island in terms of its potential behavior. At one time this stretch of beach was cut off by Hog Inlet on the north and by another inlet about 1.5 miles south of Hog Inlet, where the pier

was located. This inlet filled and was closed naturally by Hurricane Gracie in 1957 (fig. 6.2). The island was backed by tidal creek and marsh, later dredged and filled to construct the finger canals west of the main street.

The bulkheads and revetments beginning at the inlet and found along portions of the back beach indicate an erosion problem, even though the published erosion rates are less than 1 foot per year. This, together with the low elevation, the removal or erosion of former dunes, the presence of finger canals, and inlet fill, indicates that beachfront properties are at high risk with respect to coastal hazards. The character of the beach here is artificially maintained through beach nourishment, and the community is faced with a future of costly and repetitive nourishment projects. Well-marked, abundant public access walks lead to the beach. Low-risk sites can be found away from the beach, inland of the tidal creek (RM 1).

North Myrtle Beach (RMs 1 and 2)
The communities of Cherry Grove Beach, Tilghman Beach, Ocean Drive Beach, Crescent Beach, Atlantic Beach, and Windy Hill make up North Myrtle Beach. For the most part these towns fall into the category of family beach resorts with motels, restaurants, piers, and amusements. Residential development is mixed in throughout and continues inland away from the beach. Beachfront property is in the flood and V-zones (see chapter 10), and locally erosion is eating away at structures at the back of the beach (fig. 6.3). The published erosion rate of less than 1 foot per year is misleading because these communities are maintaining shoreline position through beach nourishment. Pre-Hugo dry-beach widths were 0 to 50 feet in this area (appendix D, ref. 28), and Hugo's overwash penetrated 200 feet inland. The primary dunes vary in presence and size, as does the setback of the buildings. If you are locating in this area, avoid beachfront property. Low-risk sites are abundant just a few blocks inland.

Atlantic Beach (fig. 6.4) is part of this same zone of more or less continuous beachfront de-

6.2. Cherry Grove shoreline. The inlet, closed by Hurricane Hazel in 1954, was located approximately where the pier is now.

velopment. Once an all-black community, Atlantic Beach existed for many years in social isolation, adjacent to but fenced against development. Now the community is viewed as a prime area for new development. The question is: How close will such development be to the retreating shoreline?

Windy Hill Beach, the last of these older communities, is also being replaced with newer for older and bigger for smaller as the condo-

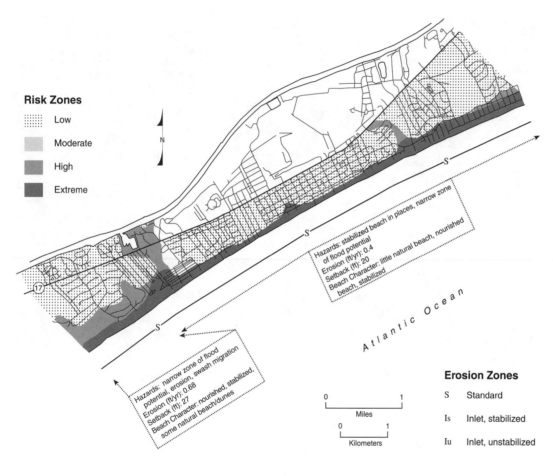

Risk Zones

Low

Moderate

High

Extreme

N

Hazards: stabilized beach in places, narrow zone
of flood potential
Erosion (ft/yr): 0.4
Setback (ft): 20
Beach Character: little natural beach, nourished
beach, stabilized

Hazards: narrow zone of flood
potential, erosion, swash migration
Erosion (ft/yr): 0.68
Setback (ft): 27
Beach Character: nourished, stabilized,
some natural beach/dunes

S

S

S

17

Atlantic Ocean

Erosion Zones

S   Standard

Is   Inlet, stabilized

Iu   Inlet, unstabilized

0   1
Miles

0   1
Kilometers

RM 2. North Myrtle Beach to White Point Swash, Myrtle Beach

miniums migrate north. The dune line here is small and offers only limited protection to beachfront houses along Ocean Boulevard.

Hurricane Hugo's storm surge was 10 to 13 feet here, but remember that in 1954 Hurricane Hazel's storm surge reached 16 to 17 feet! All the beachfront buildings in this area were destroyed or severely damaged by Hazel, and all the buildings were damaged by Hugo. The city of North Myrtle Beach initiated a $1.8 million beach nourishment project immediately after Hugo and placed 370,000 cubic yards of sand on community beaches. The sand was taken from a large shoal in Hog Inlet which had not yet stabilized after the storm, and its removal may have contributed to a later erosion problem along Hog Inlet.

White Point Swash to Singleton Swash
(RMs 2 and 3)

The south edge of Windy Hill Beach (48th Avenue) is a good place to view contrasts in land use and development. To the north runs a continuous series of communities, all urbanized to suburbanized shore with little vestige of the original environment remaining. To the south is a wider, nearly flat, truly grand strand backed by a natural buffer of dunes and forest. This area has very light development and little threatened property, but is enjoyed by thousands of campers. The area includes the quiet, and well set back, community of Briarcliff Acres. Also included in this beautiful pristine area is the religious center of Avatar Meher

Baba, whose last words were, "Don't worry, be happy." Still farther south are the high-rises and hotels—Arcadians, Hiltons, and Howard Johnsons.

The line of lakes (Sand Hill, Forked, House, Chapin, Alligator, and Long Ponds) mark a back-barrier lagoon that filled in when a barrier island welded onto the mainland. Swashes are natural drainage lines that may mark former inlet positions. The low ground makes for a wider flood zone, and the swash mouths show evidence of some lateral migration. D. K. Hubbard and coauthors (appendix D, ref. 118) reported long-term erosion on this coast, although the recorded amount varies from place to place. Caution is in order here.

Most people locating permanently in this area will be considering a condominium in a high-rise. The position of some of these structures immediately adjacent to the beach and with little or no protection could very well spell trouble in the future (fig. 6.1). Do not let the size of the building give you a false sense of security. Review the site evaluation checklist and suggestions presented in chapter 5.

Myrtle Beach Area (Singleton Swash through Lakewood; RMs 3, 4, and 5)
Myrtle Beach extends approximately 9 miles from Singleton Swash to Midway Swash, at Myrtle Beach State Park, at its southern boundary. The city developed rapidly after

6.3. Left: North Myrtle Beach is a mixture of cottages and condominiums. Structures located at the edge of the beach without natural protection are at risk. Perpetual beach replenishment will be required to maintain the recreational value of these beaches.

6.4. Right: Landward view from the beach at Atlantic Beach, a small section of the Grand Strand that has retained its original small-town character. Streets ending on the beach provide paths for storm overwash into the community.

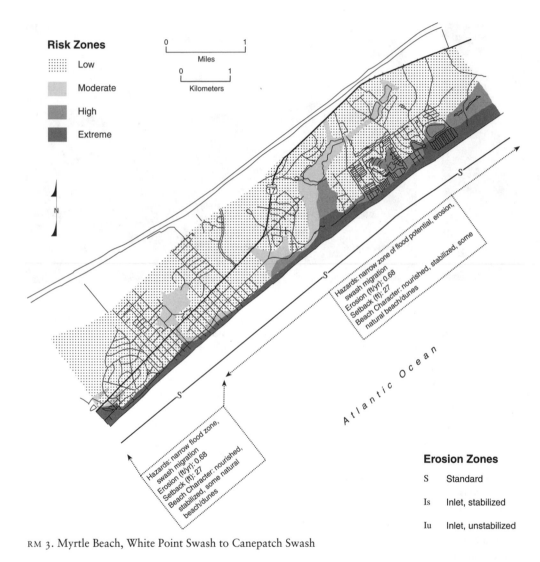

Risk Zones

:::: Low

░░ Moderate

▓▓ High

██ Extreme

0 ——————— 1
Miles

0 ——————— 1
Kilometers

N

Atlantic Ocean

Hazards: narrow zone of flood potential, erosion, swash migration
Erosion (ft/yr): 0.68
Setback (ft): 27
Beach Character: nourished, stabilized, some natural beach/dunes

Hazards: narrow flood zone, swash migration
Erosion (ft/yr): 0.68
Setback (ft): 27
Beach Character: nourished, stabilized, some natural beach/dunes

Erosion Zones

S   Standard

Is   Inlet, stabilized

Iu   Inlet, unstabilized

RM 3. Myrtle Beach, White Point Swash to Canepatch Swash

World War II and is now South Carolina's most popular seaside resort and the most active resort area between Virginia and Florida. In 1978 approximately 6.5 million people visited the Grand Strand. In 1994 the number was 12 million. The number of permanent residents is growing as well. The 1992 population of Myrtle Beach was 27,662, up nearly 3,000 from 1990.

Of all the South Carolina coast, this area is most suitable for development because the mainland coastal zone is sufficiently elevated to be above the flood zone (fig. 6.5), except for the immediate beachfront. The coast is relatively stable because it is not a migrating bar-

6.5. Public beach access site at the northern end of the city of Myrtle Beach. Notice how rapidly the elevation rises going inland. This is why large portions of Myrtle Beach are at less risk than land in North Myrtle Beach.

rier island, but erosion combined with the perception that oceanfront buildings must be protected will require either a community commitment to beach nourishment or a change in philosophy and a systematic plan for relocating buildings as they are threatened in the future. The reservoir of coarse sands in older barrier ridge deposits provides a local sand source, and local surface deposits of solid rock (e.g., at Hurl Rocks Park) add to the strand's likelihood of a low erosion rate. The lack of inlets reduces the complexity of offshore currents, so unpredictable shifts in erosion or deposition patterns are less likely. Evacuation is not a problem if people heed storm warnings.

Away from the beach, the general risk factor is low, and excellent sites abound just a few blocks inland from Ocean Boulevard. Keep in mind that there is a flood zone and a wave attack zone as well, although they are restricted to a narrower area at the back of the beach than is the case on the remainder of the coast. Avoid beachfront property. Low-elevation sites adjacent to swashes, creeks, and former marsh flats form flood corridors extending inland. These occur at 24th Avenue North, where the flood zone extends three blocks inland to Kings Highway; at 14th Avenue North, inland beyond Ocean Boulevard; all along Withers Swash and Midway Swash; in an area parallel to the oceanfront along much of Youpon Drive; and beyond Rosemary Street,

Erosion Zones

S   Standard

Is   Inlet, stabilized

Iu   Inlet, unstabilized

Hazards: narrow flood zone, swash migration
Erosion (ft/yr): 0.68
Setback (ft): 27
Beach Character: nourished, stabilized, some natural beach/dunes

Atlantic Ocean

Risk Zones

Low

Moderate

High

Extreme

0   1
Miles

0   1
Kilometers

RM 4. Myrtle Beach, Canepatch Swash to Midway Swash

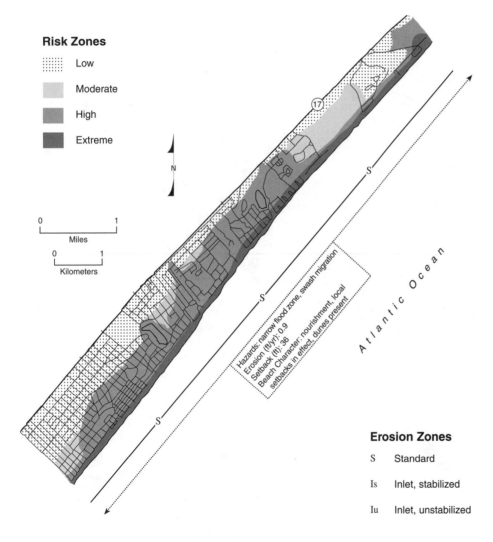

Risk Zones

:::::: Low

▨ Moderate

▨ High

▨ Extreme

N

0   1
Miles

0   1
Kilometers

17

Atlantic Ocean

Hazards: narrow flood zone, swash migration
Erosion (ft/yr): 0.9
Setback (ft): 36
Beach Character: nourishment, local
setbacks in effect, dunes present

Erosion Zones

S   Standard

Is   Inlet, stabilized

Iu   Inlet, unstabilized

RM 5. Midway Swash to Surfside to Garden City

about three blocks inland from 24th Avenue South.

The economy of Myrtle Beach is tied to medium- and high-rise resort motels, amusements, and related businesses (fig. 6.6). Much of this development faces a questionable short-term future. Such large-scale development did not exist here in 1954 when Hurricane Hazel caused so much destruction. When Hurricane Hugo struck in 1989, property loss was severe, and post-Hugo reconstruction and growth have set the stage for even greater losses in the future.

Erosion is not absent at Myrtle Beach. As is the case elsewhere along the coast, erosion must be expected because the sea level is rising. The erosion rate may not be as spectacular as on the barrier islands, but this shoreline does show an erosional trend. Erosion slowed or stabilized for a time between 1958 and 1973, a period of heavy development pressure when large buildings were constructed adjacent to the beach in the flood zone. Then erosion accelerated. In 1979 Hurricane David caused significant erosion at Myrtle Beach. In 1980 two northeasters left a scarped beach, exposed manhole housings formerly buried at the back of the beach, and caused similar minor damage. Beach scraping, the pushing of foreshore sand to the back of the beach, became a fairly common practice after storms. By 1983 most of Myrtle Beach's shorefront was totally armored with walls, revetments, and bulkheads. The beach was becoming a slim strand of its

former grand self, not the "world's widest beach" touted on local postcards.

In the 1980s Myrtle Beach residents began to recognize the significance of the erosion problem. Frequent newspaper articles, actions by local planners, support for the formation of a state chapter of the American Shore and Beach Preservation Association, Sea Grant studies, Army Corps of Engineers hearings, and similar events intensified the sense of alarm. The community opted for soft stabilization through beach nourishment and locally adopted setback regulations in an effort to both protect property and work with nature to bring back the wide, unobstructed beach. A beach fill project was completed in the mid-

1980s with the emplacement of 652,430 cubic meters of sand. By May 1988, 460,000 cubic meters of the nourishment remained on the beach, according to the OCRM's annual "State of the Beach" report. The city also began a program to eliminate storm-water outfalls on the beach, and the state established setback requirements in anticipation of future shoreline regulation.

The artificial beach fill offered some protection from Hurricane Hugo (appendix D, refs. 28, 30), but severe erosion of the beach, marked by scarping, removed the fill and damaged or destroyed seawalls. Thirty-seven of the 74 survey monuments installed by Coastal Science and Engineering, Inc., a Columbia-based

6.6. Left: Withers Swash on Myrtle Beach. The swash (small channel) runs from the small pond in the upper right and empties into the sea just to the south (left) of the roller coaster. Myrtle Beach is a mainland coast without barrier islands. The swashes provide a means of freshwater and saltwater drainage to the sea.

6.7. Right: Myrtle Beach State Park represents a coastal land use that maximizes recreational beach use without placing property at risk. Photograph courtesy of Kay Carlson.

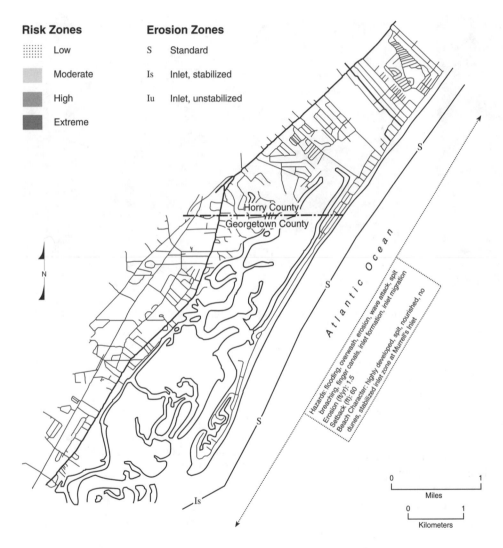

Risk Zones

▦	Low
▨	Moderate
▨	High
▪	Extreme

Erosion Zones

S	Standard
Is	Inlet, stabilized
Iu	Inlet, unstabilized

Horry County
Georgetown County

Atlantic Ocean

Hazards: flooding, overwash, erosion, wave attack, spit breaching, finger canals, inlet formation, inlet migration
Erosion (ft/yr): 1.5
Setback (ft): 60
Beach Character: highly developed spit, nourished, no dunes, stabilized inlet zone at Murrell's Inlet

N

0   Miles   1

0   Kilometers   1

RM 6. Garden City

consulting firm, to monitor the behavior of the beach fill project were lost or destroyed by Hugo-related erosion. High-water marks inside buildings along the shoreline were 12 to 15 feet above sea level (appendix D, ref. 32). When the shoreline is blocked by medium- and high-rise buildings, the storm-surge flood and ebb currents are magnified, concentrating overwash between buildings, over parking lots, and along shore-perpendicular streets. The return ebb flow scoured channels up to 3 feet deep and as much as 200 feet wide across the beach and offshore, and carried debris of all types more than 300 feet offshore to depths of 12 feet, creating offshore obstructions (appendix D, ref. 27).

After Hugo, 8.5 miles of the city's beach needed emergency renourishment. Showing foresight, the city had purchased a sand pit located inland, and 380,000 cubic yards of sand were placed on the beach. More than 100,000 truckloads of sand were taken from the spot to replenish local beaches. Eventually, however, the sand source was depleted and local residents objected to the frequent truck traffic. By 1995, Myrtle Beach had made plans to go offshore for future beach nourishment sand supply. Clearly, if Myrtle Beach is to keep its beach—its *raison d'être*—it must make an eternal commitment to beach nourishment. Who will pay for this? The taxpayers.

From Myrtle Beach State Park to Lakewood (north side of Surfside Beach), development is light. The state park (fig. 6.7) has preserved much of the forest as a campground, and the nature trail provides a glimpse of what the environment was like before it was covered by concrete, asphalt, and roller coasters. The popular park also sports a fishing pier, swimming pool, and cabins. The flood zone extends farther inland around Lewis Lake and its tributaries, but other inland areas are in the low-risk zone.

Surfside Beach (RM 5)

Surfside Beach (Floral Beach on older maps) is a residential development dating back to the 1950s. Although the average erosion rate here is less than 1 foot per year, the city's setback ordinance (which preceded the state's) provides a buffer zone between beach and buildings. The setback is most likely responsible for the higher property values and condominium occupancy here (appendix D, ref. 30). It also may have reduced the impact of Hurricane Hugo. Nevertheless, the storm caused 28 feet of beach recession (appendix D, ref. 31), and an emergency beach nourishment project placed more than 70,000 cubic yards of sand on 1.3 miles of Surfside Beach. The best building sites are those away from the beachfront, out of the flood zone in the western part of the community (toward U.S. 17).

Garden City (Horry County portion; RM 6)

Garden City (fig. 6.8) is the southernmost coastal community in Horry County. The changing character of the coast—from main-

land shore to barrier island–like spits adjacent to inlets and backed by marsh—increases the risk factor here. The inlet hazard is greater, the flood zone extends progressively farther inland, and the measured erosion rates tend to increase in this area (1–3 feet/year). The beachfront can be characterized as subject to moderate erosion, but recession in any given year can be as much as 3 to 15 feet.

6.8. Storm damage at Garden City after Hugo. The seawall shown here was also severely damaged in the January 1, 1987, northeaster. The wall and pools were repaired but were destroyed by Hurricane Hugo.

Front beach property (fig. 6.8) should be avoided. There are a few elevated sites on a frontal ridge, but the addition of wave height to the 100-year flood level probably places these locations in the flood zone as well. The best sites are those on the inland side of the marsh at elevations above the flood zone. Keep in mind that hurricane winds are a threat even here, and take the necessary steps in construction to improve wind resistance.

Georgetown County

From north to south the coastline of Georgetown County goes through the transition from mainland shore to barrier islands. Beginning at Garden City with the spit north of Murrell's Inlet and proceeding down the Waccamaw Neck (the land between the Waccamaw River and the sea), distinct barrier forms become apparent. Of these, the best known is Pawleys Island. Distinct islands lie off the mouth of Winyah Bay (North Island, South Island, and Cat Island), giving way to the Santee Delta and Cape Romain. The hazard potential rises with the increasing number of inlets, with lower elevations that place larger areas in the flood zone, and where shoreline erosion and overwash are more common. The average erosion rate on Debordieu Island is 11.5 feet per year! Beachfront property and sites on spits adjacent to inlets are generally at extreme risk, but there are some sites at lower risk (moderate to high) because

of their elevation, dune protection, vegetation, and location away from inlets or erosion zones, as noted below.

Garden City (Georgetown County portion; RM 6)

This narrow spit has seen significant development in the 1990s. Because of the low elevations, flood potential, and moderate erosion rates (1–3 feet/year), the area is at extreme risk from the county line to the area of the finger canals. The potentially destabilizing effect of the finger canals and the proximity of Murrell's Inlet also place the southern extent of the spit in the extreme-risk category. Murrell's Inlet is stabilized now, and the jetty system has a limited sand-trapping effect that causes accretion (beach buildup). Such accretion should offer additional protection to property, but this advantage is offset by potentially rapid erosion during storms, or perhaps even inlet formation (breaching) north of the stabilized inlet. The depositional basin of the Murrell's Inlet jetties was the borrow site for post-Hugo beach nourishment at Garden City.

Numerous condominiums were built along the shoreline here, with bulkheads placed in front of structures right on the beach. The beach was becoming narrower and degrading even before Hugo. The existence of such hardening structures is a clue to property owners to expect continuing costs for bulkhead maintenance and repair.

In contrast to Surfside, a community with a setback requirement, South Garden City, with no setback and an armored shoreline, experienced beach recession of 42 feet (surfside lost 28 feet) and severe damage to all shorefront houses from Hugo; the storm inundated the spit and eroded the dunes away. Houses were damaged three blocks inland, and the poorly built shore protection structures were no match for the storm (appendix D, ref. 29). In the area of Kingfisher Pier, debris was carried into the marsh, some 800 feet inland. Some houses were transported inland up to 400 feet along with the overwash (see fig. 5.6). Paul Gayes of the University of South Carolina, Coastal Carolina College, identified offshore topographic features (appendix D, ref. 27) that correlated with gaps between buildings. The gaps funneled the returning ebb flow through the nearly solid front of shore buildings. Ebb-surge scouring and channeling were enhanced by these gaps and by roads and paths running perpendicular to the shore. Without doubt, such currents carried sand offshore. Garden City required more than 163,000 cubic yards of beach nourishment along nearly 2.5 miles of beach after Hurricane Hugo.

Huntington Beach State Park (RM 7)
Although Huntington Beach State Park will not be developed beyond the level of park use, the risk factor may be illustrated as if development were being considered. Adjacent to the inlet is an extreme-risk zone, which decreases

to high risk along the park's beachfront. Erosion rates in this area have fluctuated considerably over the years and have been severe at times. Ponds were formed in low areas at the back of the beach as a result of Hugo, and drainage from the ponds created temporary breaches. The park was renourished twice prior to 1995 by the U.S. Army Corps of Engineers using sand trapped by the updrift jetty system of Murrell's Inlet. Such nourishment probably benefits downdrift beaches.

North Litchfield Beach (Retreat Beach; RM 7)
This short segment of coast is mainland without an intervening marsh zone. The shoreline is relatively stable, and storm-surge flooding and wind are the greatest threats. Sites range from extreme and high risk in the vicinity of the shore to low risk in the more inland positions west of U.S. 17.

Litchfield Beach (Magnolia Beach; RM 8)
This variable-risk zone includes Litchfield and Litchfield-by-the Sea. The ever-present flood potential, local finger canals, and variable erosion patterns make it necessary to evaluate each site carefully. The access road is at low elevation in the vicinity of the marsh. An artificial bulldozed dune ridge, approximately 20 feet high, survived Hugo to some degree and protected development, although flooding occurred from the landward side. However, the total area of the dune field was significantly reduced by the storm (appendix D, ref. 30). The

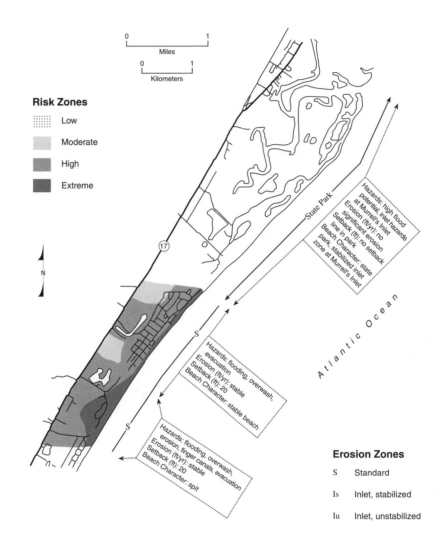

Risk Zones

- Low
- Moderate
- High
- Extreme

Erosion Zones

S   Standard

Is   Inlet, stabilized

Iu   Inlet, unstabilized

Hazards: high flood potential, inlet hazards at Murrell's Inlet
Erosion (ft/yr): no significant erosion
Setback (ft): no setback line in park
Beach Character: state park, stabilized inlet zone at Murrell's Inlet

Hazards: flooding, overwash, evacuation
Erosion (ft/yr): stable
Setback (ft): 20
Beach Character: stable beach

Hazards: flooding, overwash, erosion, finger canals, evacuation
Erosion (ft/yr): stable
Setback (ft): 20
Beach Character: spit

RM 7. Huntington Beach State Park, North Litchfield, and Litchfield Beach

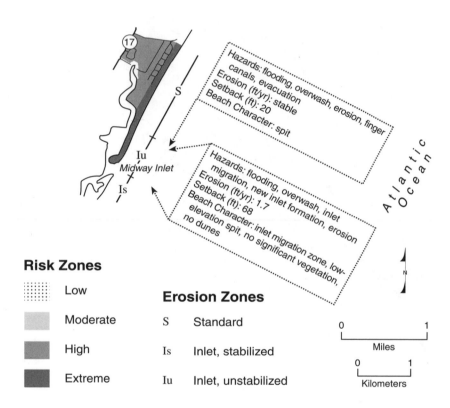

Hazards: flooding, overwash, erosion, finger
canals, evacuation
Erosion (ft/yr): stable
Setback (ft): 20
Beach Character: spit

Hazards: flooding, overwash, inlet
migration, new inlet formation, erosion
Erosion (ft/yr): 68
Setback (ft): 1.7
Beach Character: inlet migration zone, low-
elevation spit, no significant vegetation,
no dunes

Atlantic Ocean

N

Risk Zones

Low

Moderate

High

Extreme

Erosion Zones

S   Standard

Is   Inlet, stabilized

Iu   Inlet, unstabilized

0 ────── 1
Miles

0 ────── 1
Kilometers

RM 8. Litchfield Beach and Midway Inlet North

preexisting bulkhead, which was buried prior to the storm, was exhumed and extensively damaged. Along another stretch of Litchfield, high dunes were completely destroyed by Hugo, but there was no development there to be affected (fig. 6.9A, B).

Litchfield-by-the-Sea (fig. 6.10) has avoided beachfront development (except for a community beach house), has preserved its dunes and forest, and has maintained distant setbacks. In contrast, Litchfield extends to the back side of a good frontal dune line, and finger canals intrude on former marsh. The Hugo experience in this area demonstrated that the best measures of frontal protection are high, wide dune fields, without gaps, fronted by wide dry-beach widths (prestorm). Communities in this zone must be committed to the continuous and growing costs of building, maintaining, and vegetating a dune system as well as beach nourishment.

Litchfield Spit (McKensie Beach; RM 8)
Ideally, development should not be allowed on Litchfield Spit (fig. 6.11), south of Litchfield Beach, but the land on the spit is privately owned and it is possible that development will take place. At the time of this writing, the OCRM has received applications for 17 dock permits for lots at the back of this spit.

Hugo's destruction of this low and vulnerable dune field upset a development plan. Like all spits and shores adjacent to inlets, this finger of sand displays the multiple risk factors

that make it an extreme-hazard zone. Elevations are too low to be out of the flood zone—or even, in many locations, out of the V-zone, where wave impact intensifies the hazard. Overwash potential is high where dunes are low, absent, or have gaps. A change in the ebb-tidal delta can cause an abrupt shift from accretion to rapid erosion. The published erosion rates for the spit are on the order of 1.7 feet per year; however, this figure describes erosion in the landward direction only, not migration of the inlet, cutting of a new inlet, or erosion of the whole spit down to sea level. Inlet migration is a certainty. New inlet formation or breaching during hurricanes was predicted,

and indeed, Hurricane Hugo's waves and storm surge eroded the entire spit down to sea level.

Access to this land is only through Litchfield Beach. Furthermore, the entire spit is in a Co-BIA zone (see chapter 10), so there is no federal subsidy for development, infrastructure, or maintenance. The property owners of this community should take a lesson from Folly Beach, which blocked development on a similar spit.

Pawleys Island (RM 9)
The development of Pawleys Island dates back to the rice planter days of the eighteenth cen-

6.9. Hurricane Hugo completely removed this natural primary, front-beach dune along Litchfield Beach. Left: Dune before Hugo. Right: Flattened dune site after Hugo. Elsewhere at Litchfield, a roughly 20-foot dune constructed following Hurricane Hazel (1954) survived Hugo and helped preserve landward structures.

6.10. Litchfield-by-the-Sea illustrates a good building setback policy. This photo was taken in 1981; today these houses and others nearby are closer to the sea but not yet threatened.

1893 and 1954 hurricanes were forgotten in less than a generation. Cottages at the edge of the beach became yardsticks by which to measure shoreline retreat, and the pressure to "stabilize" the beach was answered in the late 1940s with the first groins. These groins were expanded or added to over the years. Single-family residences were replaced by condominiums, crowding even more people into the same limited area. The addition of a municipal waste treatment system, pressure to rezone from residential to commercial, suggestions to add additional hard stabilization structures, and the change in status of a fishing pier from public to private were all signs that the development process was not over. An island described as having an air of quiet pleasurability was drifting toward the din of commercialism, all in the face of extreme- and high-risk coastal hazard conditions. Incorporation in 1985 was meant to check the commercialism, but the post-Hugo reconstruction has seen an ever greater concentration of property value on the island as the destroyed island cottages were replaced by $500,000 beachfront houses on lots that sold for $350,000.

Pawleys Island has 3.5 miles of beachfront, is 0.5 mile wide, and is bounded by inlets, all of which suggest that significant changes

tury, and its promotional literature bills the island as "the oldest resort in America." The island's history could serve as a case study for what is likely to happen on almost all developed barrier islands. Causeway access opened the doors for development. Large lots were divided and subdivided again and again, and today's development is a warren of large beach houses. The development history is punctuated by the island's storm history. Weak structures on high-risk sites such as the beachfront and the spit have been destroyed, whereas well-crafted nineteenth-century buildings placed on high ground behind dunes under a canopy of protective vegetation have survived. Each time storms destroyed the high-risk buildings, the buildings were replaced by even more precariously positioned structures. The lessons of the

6.11. Litchfield Beach and spit. The spit represents an extreme-risk zone in a class 3 hurricane. It was completely overwashed during Hugo, but pressure to develop the spit continues.

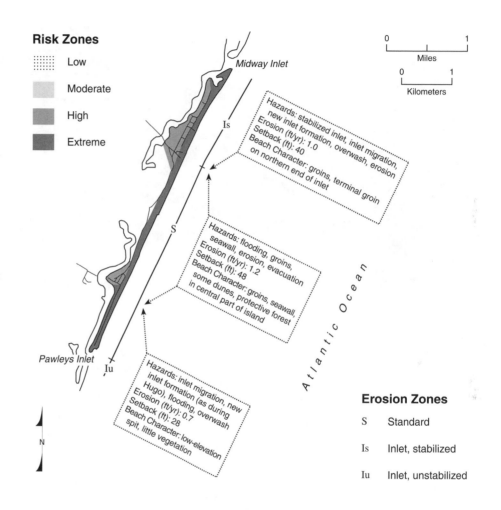

Risk Zones

::::: Low

▢ Moderate

▨ High

■ Extreme

Midway Inlet

Is

Hazards: stabilized inlet, inlet migration, new inlet formation, overwash, erosion
Erosion (ft/yr): 1.0
Setback (ft): 40
Beach Character: groins, terminal groin on northern end of inlet

S

Hazards: flooding, groins, seawall, erosion, evacuation
Erosion (ft/yr): 1.2
Setback (ft): 48
Beach Character: groins, seawall, some dunes, protective forest in central part of island

Pawleys Inlet

Iu

Hazards: inlet migration, new inlet formation, flooding (as during Hugo), flooding, overwash
Erosion (ft/yr): 0.7
Setback (ft): 28
Beach Character: low-elevation spit, little vegetation

Atlantic Ocean

N

0   1
Miles

0   1
Kilometers

Erosion Zones

S   Standard

Is   Inlet, stabilized

Iu   Inlet, unstabilized

RM 9. Midway Inlet South, Pawleys Island to Pawleys Inlet

Individual Island Analyses   **91**

6.12. Left: A lesson from Hurricane Hugo on Pawleys Island. This house with protective elevation, dunes, and vegetation survived with little damage.

6.13. Right: This new inlet formed by Hurricane Hugo through Pawleys Island spit had to be filled artificially in order to restore access. Compare this photo with fig. 1.6. Post-Hugo reconstruction has placed bigger housing units in the same extreme-risk zone. Perhaps the spit should not have been redeveloped.

should be expected. The groin field attests to the erosion and in places may contribute to it. Shoreline trends vary from a few places with short-term accretion to places with erosion rates as high as 7 feet per year. However, most of the island front eroded at a rate of 2 feet per year between 1872 and 1966 (appendix D, refs. 46, 47); more recently published rates are 1.2 feet per year. Protective dunes, vegetation, and elevation vary, but where the three are favorable to lower-risk construction you will usually find one of the older buildings (fig. 6.12).

During Hurricane Hugo, the large dunes blocked overwash, and forested areas protected houses from wind and flooding. Low dunes also survived, but only because they were rapidly flooded and overtopped by the rising sea (appendix D, ref. 30). Dune fronts did their natural job, yielding nourishing sand to the beach as they were eroded. New dune gaps formed locally to allow overwash to penetrate the island and storm waters to attack buildings behind the gaps. Buildings with no dunes or only low dunes for protection were destroyed or heavily damaged, as were bulkheads and low seawalls, which were easily overtopped and afforded little protection to structures behind them. More than 3 miles of shoreline required emergency beach nourishment after the storm. This renourishment was controversial because it required a beachfront management plan to be filed with the OCRM. Pawleys Island officials were reluctant to do that because the community provides poor public access to the beach. According to a March 9, 1991, article in the *Sun News,* Pawleys officials finally agreed to prepare a beachfront management plan but had made no plans at that time to improve public access.

The north end of Pawleys Island is an extreme-risk zone because of the dynamic, now partially stabilized, Midway Inlet (see fig. 4.6), which migrated 1,300 feet south between 1939 and 1950 (appendix D, ref. 70), and the potential for deep penetration of overwash or even inlet breaching along shore-perpendicular streets that run from the ocean side to the marsh side of the island (e.g., Shell Drive and 2nd and 3rd Streets). Storm-wave attack, overwash, and flooding are near certainties during even a small hurricane. In 1979 Hurricane David severely eroded the north end of the island, and later northeasters have caused beach scarping, resulting in minor damage to beachfront property.

The risk is high to moderate from the middle portion of the island south to the wider, forested part. This area includes the low-risk sites noted above, but there are also gaps in the dunes where you can stand on the beach and look directly at the fronts of buildings. These openings will be overwash and flood passes just as they were during Hurricane Hugo; avoid them. Avoid low elevations on the back side of the island as well because floodwaters

from storms will rise out of the marsh. Beachfront property near groins may experience sand starvation and accelerated erosion.

The flood levels in excess of 13 feet during Hurricane Hugo should not have come as a surprise. During Hurricane Hazel in 1954 the high tide is estimated to have risen to 11.5 feet above mean sea level at Pawleys Island, topped by waves with an estimated height of 10 feet. All of the island was affected, but nowhere more severely than the south spit end, where 27 of 29 houses were destroyed and a wide breach channel was eroded across the spit. The spit is an example of an extreme-risk zone that should never have been developed. Certainly it should not have been redeveloped after the storm, but redeveloped it was. Unfortunately, the setbacks required by the Beachfront Management Act do not address this kind of shore-parallel erosion. Where a row of houses had been destroyed, another row of bigger houses was soon facing the open sea without any protection. In the first edition of *Living with the South Carolina Shore* (appendix D, ref. 58) we noted that this end of the island was accreting but predicted that there was insufficient time for protective dunes to form before the next hurricane. These houses were thus in both the flood zone and the wave velocity zone, and the spit was likely to be breached again. Hurricane Hugo did just that, completely flooding the spit, eroding a small inlet (fig. 6.13), and either totally destroying or severely damaging the houses. But history repeats itself. The breach

was filled and the houses rebuilt—bigger, better, and more expensive ones! Another prediction: These houses will meet the same fate as their predecessors, possibly before their mortgages are paid off.

If you live anywhere on this island, evacuate early. The distance to a safe haven inland may be short, but the low elevations along the escape route may flood in advance of a hurricane's landfall. People who stay on barrier islands during hurricanes may be like the gray man; they may disappear.

Debordieu Island (also appears on maps as Dubourdieu Island, Dubordie Island, and Debidue Beach; RM 10)

Debordieu Island (fig. 6.14), the southern extension of Waccamaw Neck, is not completely separated from the neck by marsh and tidal creek drainage, but this coastal strip is a barrier island, and the processes occurring here and the corresponding hazards are those of an island. Three sections can be recognized.

The northern portion of the island is undeveloped, and with good reason. Pawleys Inlet has shown a progressive migration to the south, eroding the north end of Debordieu. The area is low, lies in the flood zone, and has been damaged in the past by storms that caused erosion and overwash. At times the beach has accreted, but the additions have never been permanent. Toward the developed part of the island there are active sand dunes.

Erosion Zones

S   Standard

Is   Inlet, stabilized

Iu   Inlet, unstabilized

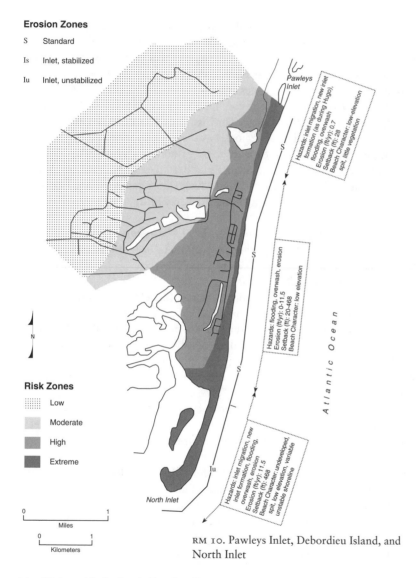

Pawleys
Inlet

Hazards: inlet migration, new inlet
formation (as during Hugo),
flooding, overwash
Erosion (ft/yr): 0.7
Setback (ft): 28
Beach Character: low-elevation
spit, little vegetation

Hazards: flooding, overwash, erosion
Erosion (ft/yr): 0-11.5
Setback (ft): 20-468
Beach Character: low elevation

A t l a n t i c   O c e a n

Risk Zones

Low

Moderate

High

Extreme

North Inlet

Iu

Hazards: inlet migration, new
inlet formation, flooding,
overwash, erosion
Erosion (ft/yr): 11.5
Setback (ft): 468
Beach Character: undeveloped
spit, low elevation, variable
unstable shoreline

N

0   1
Miles

0   1
Kilometers

RM 10. Pawleys Inlet, Debordieu Island, and
North Inlet

Apparently there were once buildings here, but they were buried by the dunes years ago. These dunes prevented overwash during Hugo. Although this land has been occupied since the early 1800s as part of the plantation system, the island remained undivided until modern subdivision and development.

The middle section includes the luxury development of Debordieu Colony and the forested high ground on the southern half of the island. This area includes some moderate-risk sites on the higher ground away from the beach, but much of the area is extreme to high risk. Most of the beachfront property should be regarded with great caution because of the erosion and overwash potential. Estimated water depths during Hugo were above 12 feet (appendix D, ref. 32), and the artificial sand dune behind the wall with a crest elevation of 17 feet and basal width of more than 95 feet was completely eroded. The houses behind the dune that were built on concrete slabs were torn apart and the debris carried inland for

6.14. Facing page: Aerial view of Debordieu Island showing narrow beach in front of seawall and wide beach where there is no wall. (A) Top: Note the maritime forest and marsh behind the beach. (B) Bottom: A small bulkhead fronts large private homes on the exclusive island, shown in a low-tide photograph. The low dune and bulkhead afford little protection against storm-surge flooding and wave action but will halt (temporarily) shoreline retreat.

hundreds of feet. One house floated off its foundation and was transported more than 80 feet until it collided with another house. Significant beach erosion occurred. Erosional scarps (vertical bluffs) were present in the front row of dunes and marsh peat was exposed on the beach in the years prior to Hugo, indicating shoreline retreat. Keep in mind that costly, exclusive development does not necessarily imply safety from hazards. Portions of this island are eroding at an average of 11.5 feet per year. Look for site elevations of 10 feet or more, fronted by protective dunes and healthy vegetation cover, plus sufficient setback to avoid the consequences of rapid erosion.

The southern extension of the island is a relatively barren spit that has grown into North Inlet. Although accretion has been rapid since the 1930s, this undeveloped, low-elevation sand flat is an extreme-risk zone subject to flooding and overwash. Its narrow width makes it likely that inlet breaching such as occurred during Hugo will occur again during future hurricanes. The spit is unsuitable for development.

The Santee Delta Coast

Georgetown County, continued

This shoreline stretches from North Inlet to Capers Inlet and includes Winyah Bay, Cape Romain, and the Santee Delta and adjoining Bulls Bay. This section of coast includes several small islands that are state and national refuges and coastal preserves (fig. 6.15). These islands are not shown on the risk maps because they will not be developed. Some are open to limited recreational use (see appendix B, "Parks and Recreation"), and all are examples of barrier island types in their evolution and their response to coastal processes.

North Island (fig. 6.15)
North Island is part of the Thomas Yawkey State Wildlife Refuge, an important migratory bird refuge. Although the island is large, much of it is marsh. The front side of the island consists of forested beach ridges with elevations up to 42 feet; however, there is evidence of rapid erosion along most of the northern part of the island, particularly the inlet shore. The beachfront is littered with trees, marsh sediment is exposed, and the beach has advanced into the edge of the forest. Overwash fans penetrate beyond the beach. In contrast, along the southern portion of the island accretion was the rule prior to Hurricane Hugo, and a sparsely vegetated, continuous dune separated the beach and the forest. Hugo completely overwashed the southern part of the island (appendix D, ref. 32).

North Island was a familiar gathering place in plantation days, and the large village of Lafayette grew at North Inlet on the north end of the Island. The church and approximately 100 houses that made up the village were swept away in the great hurricane of 1822.

6.15. Index map of the Santee River Delta islands.

Georgetown

North Island

North Santee River

Cat Island

South Santee River

South Island

Cedar Island

Murphy Island

Raccoon Key

Cape Island

Bulls Bay

Bull Island

Capers Island

Dewees Island

0   10 miles

0   10 kilometers

Undaunted, residents rebuilt the resort, which remained popular through the 1820s and 1830s. Ultimately it fell victim to later hurricanes and competition from newer villages with easier access.

The location of the lost village of Lafayette next to North Inlet says something about inlet dynamics. The erosion along the northern section of the island is probably related to migration of North Inlet. The highest erosion rates in the vicinity of North Inlet are on the order of 11.5 feet per year. Longshore drift is to the south, and the north jetty at the mouth of Winyah Bay may be contributing to accretion along the southern shore.

South Island (fig. 6.15)

South Island is also part of the Thomas Yawkey Wildlife Refuge. This large island lies between Winyah Bay and North Santee Bay, consists of extensive marsh, and is separated from the marsh of Cat Island by a small tidal creek. On the oceanfront, South Island has 5.5 miles of beach backed by forested beach ridges that have a maximum elevation of 21 feet. A few Santee planters had summer homes on the island in the early 1800s, but the island will now remain undeveloped.

Cedar Island (figs. 6.15, 6.16)

Cedar Island lies between the North and South Santee Rivers. Its southern extremity faces the ocean and has the character of a barrier island spit. The 3-mile-long beachfront is under the

influence of the river mouths and experiences significant erosion. The inner marshland of the island was once used to grow rice, and at one time there was a small village on the island. The island is now part of the Santee Coastal Reserve and is available for limited recreational use by permit from the state.

6.16. Cedar Island on the Santee River Delta. Note the sandbars on the associated ebb-tidal delta. Such deltas represent a natural reservoir of sand for adjacent barrier islands. There was once a small town on Cedar Island, long since lost to the forces of the sea.

Charleston County (RMs 11, 12, 13, and 14)

Charleston County includes Cape Romain and part of the great Santee Delta in the north and extends to Edisto Beach in the south. This area includes some of the best in coastal development, such as Kiawah Island and the high elevations of the Isle of Palms. Unfortunately, it also includes some very high-risk areas like Dewees Island and Edingsville Beach.

Range in development type is not the only diversity that characterizes the coast of Charleston County. The county encompasses the transition zone between the microtidal, wave-dominated coast to the north and the mesotidal shoreline of the low country. It has both narrow, retreating barrier islands like Morris Island and wide beach ridge islands like Kiawah. The inlets also range in size, from small Jeremy Inlet north of Edisto Island to medium-sized Captain Sams Inlet and giant flooded river valleys like Charleston Harbor and the North Edisto. Erosion rates range from highly erosional to moderately accretional. Charleston County has both the most pristine shores and the most engineered shoreline on the southeastern seaboard. Diversity rules here—Charleston County has it all.

Murphy Island (fig. 6.15)
Murphy Island is part of the Santee Coastal Preserve and is managed by the state. The island will not be developed, although limited recreational use is permitted. The 6-mile curv-

ing beachfront has shifted significantly through time, accreting as the delta grew, retreating as the delta's sediment supply was interrupted or the Santee River shifted its course. Although the island is mostly marsh, there is a small area of dune ridges. As part of the Harry Plantation the island was used for rice cultivation prior to 1836, and slave villages were located here. After slaves were lost to hurricanes, so-called hurricane shelters were built as storm refuges. These refuges were not particularly effective, given the low elevation of the island, but they were at least an alternative in the days when evacuation was impossible.

Cape Island (fig. 6.15)
Cape Island is part of the Cape Romain National Wildlife Refuge, which means that this island remains in its natural state and will not be developed. The island forms the apex of Cape Romain and is eroding rapidly along most of its beachfront. Sand transported to the ends of the island is accreting to form the spits. Elevations are generally less than 10 feet above sea level, and the island is not suitable for development. The island was breached by Hurricane Hugo in 1989.

Lighthouse Island and Raccoon Key (fig. 6.15)
Lighthouse Island and Raccoon Key are two of several marsh islands that lie behind Cape Island and are part of the national wildlife refuge. Lighthouse Island and Raccoon Key both have ocean beachfront, and neither is suitable

for development. Raccoon Key in particular is very unstable; its elevations of less than 5 feet mean that it is frequently overwashed. Erosion rates have been measured in hundreds of feet per year, and the beaches are floored by exposed, eroding marsh muds. Three breaches were opened by Hugo across Lighthouse Island, and one across Sandy Point Beach.

Bulls Bay (fig. 6.15)
Bulls Bay forms a large reentrant into the mainland coast between the barriers of Cape Island and Bull Island. The bay's shores are protected marshlands that make up the Cape Romain National Wildlife Refuge, which in turn is backed by the Intracoastal Waterway. You might assume that no shoreline development means no problems from storms, but the area provides an example of why you should consider the coastal zone as much as the shoreline when evaluating hazards. The bay and its associated islands, which fortunately are uninhabited, were to the right of Hurricane Hugo's eye and were exposed to the storm's maximum sustained wind velocity, estimated at 135 miles per hour.

McClellanville (fig. 6.15)
McClellanville lies about 6 miles inland from Cape Romain behind miles of marshland. Prior to Hurricane Hugo, the residents thought they were safe from storm surge, but distance did little to mitigate Hugo's storm surge of 18 feet. The entire distance from Cape Romain al-

6.17. The general state of affairs on the mainland in McClellanville after Hugo.

placed them in the ceiling ducts. Finally, at 3:00 A.M., the water level fell, and a few hours later the evacuees emerged into a scene of total devastation. McClellanville learned three important lessons: (1) evacuate outside the coastal zone in the event of a hurricane, (2) do not base your actions on past hurricane behavior or a perception that you are experienced because you survived past storms, and (3) elevation well above potential storm-surge flood levels should be a top priority in choosing inland evacuation centers or sites of refuge.

Bulls Island (fig. 6.16)

Bulls (or Bull) Island is bounded on the north by Bulls Bay and on the south by Price Inlet. The shoreline of Bulls Island is affected by a complex pattern of wave refraction. Maps show this coast to be unstable and subject to long-term erosion (fig. 5.2), but this is of little consequence to the nesting birds and loggerhead turtles that are its most important users. The island is part of the Cape Romain National Wildlife Refuge. Numerous trails crisscross the island, and hikers and birdwatchers can enjoy the forested dune ridges or walk along the 6.5 miles of beach.

Capers Island (fig. 6.15)

Capers Island, also called Big Capers Island to differentiate it from an island of the same

most to U.S. 17 is nothing but very low elevation salt marsh. As Hugo approached, some residents left, but many stayed in their homes or went to the designated evacuation center, Lincoln High School. As the water level rose rapidly between midnight and 1:30 A.M., residents who had remained in their homes were forced to the second floors, which proved safer from flooding than the designated evacuation center (appendix D, ref. 32). By 2:00 A.M. the

water level in the school was chest deep, and the deafening noise outside, which those who heard it compared to roaring locomotives, added to the terror of the night. Outside, the surge was transporting the debris from trees, docks, and buildings; tearing the bark off trees; and destroying the town's fishing fleet and virtually every vehicle in the town (fig. 6.17). The water continued to rise in the school, and in near-total darkness, adults lifted children and name in Beaufort County, is part of the South Carolina Heritage Trust. Set aside as a heritage preserve, the island is under the management of the state Wildlife and Marine Resources Department and will not be developed.

Capers Island is a natural laboratory in which to observe the influence of adjacent inlets—Price and Capers—on a small island (3.3 miles). Although it was part of the eighteenth-century plantation system and was used for a bit of agriculture before World War II, the front of the island has undergone almost continuous retreat since map records have been kept. The island lost an average of 15 feet of beachfront per year over 107 years of record. The beach is an obstacle course of fallen trees and trees standing upright where the waves have torn away the sand footing, and overwash penetrates the island front, all part of the island's natural migration. Dune ridges have been truncated and scarped. Capers Inlet on the south side has migrated first to the south, eroding Dewees Island, and then back to the north in recent years, eating back into Capers Island.

The Central Coast

The central coast of South Carolina extends south from Capers Inlet through some of the state's more famous barrier islands to Edisto Beach. Most of the islands are of the drumstick variety, short, and separated by numerous small inlets (appendix D, refs. 46, 47).

Charleston Harbor is the largest estuary along the county's shore and is essential to the region's economy. Overall, the county's shoreline is dominated by erosional trends (appendix D, ref. 118). Erosion and overwash are severe on some of the low, narrow islands. Hurricanes have flooded all these islands in the past, sometimes with loss of lives and extensive loss of property. Man-made structures such as the Charleston Harbor jetties and island groin fields have added to the erosion problem. Locally, hurricane evacuation routes may not be adequate. Fortunately, many people heeded the warnings for Hugo and evacuated early, no doubt saving lives. Most of this coast must be classified as high to extreme risk. Beachfront property should be avoided.

From north to south the islands are Dewees Island, Isle of Palms, Sullivans Island, Morris Island, Folly Island, Oak Island, Long Island, Kiawah Island, Seabrook Island, Deveaux Bank, Botany Bay Island, Botany Plantation (Botany Island), and Edingsville Beach.

Charleston County, continued

Dewees Island (fig. 6.15)
Dewees Island has only 2.9 miles of beachfront. As noted above, it is at times eroded by Capers Inlet, and similarly by Dewees Inlet along its southern margin. The island is so small that changes in either inlet and the associated ebb-tidal deltas may drastically change patterns of erosion or accretion. No

risk map was made for Dewees Island, and none is needed; its risk classification is very simple. The entire island is in a V-zone and is thus extremely hazardous. Although the central portion of the island includes a small area of forested dune ridges, no part of it is suitable for development. The immediate shorefront is mapped as erosional to unstable (appendix D, ref. 118), and between 1941 and 1973 the central and southern portions of the island receded more than 900 feet.

Adding to the extreme-risk rating of Dewees is its inherent instability, a result of its small size and the adjacent inlets. The island is accessible only by boat, and it is not likely that a bridge will ever be built. The state holds an easement that prevents development on the northern side of the island. In spite of these very obvious hazards, Dewees Island is being developed as an exclusive, high-priced community. The high price will become evident when the next Hugo roars through and we taxpayers have to foot the bill for cleanup and repair.

Isle of Palms (RM 11)
Isle of Palms is a 6.2-mile-long island suburb of Charleston with the classic drumstick shape; it is bounded and influenced by inlets. The northern section is under the strong influence of Dewees Inlet and shows alternating trends of accretion and erosion. Between 1958 and 1968 parts of this section lost up to 600 feet of beachfront (appendix D, ref. 118), and erosion also occurred during Hugo. There is higher

forested ground inland from the beach where lower-risk sites abound, and this section may be viewed as being at high to moderate risk depending on the character of the individual site. The highest elevations are in the area of the Wild Dunes development on the northern third of the island. Unfortunately, bulldozing lowered some elevation and removed protective forest cover during development, increasing the hazard potential. The Wild Dunes Beach and Racquet Club has its golf course located nearest the inlet, a better land use than locating buildings in the high-hazard zone. By the same token, however, some holes of the golf course are on the highest, most heavily forested parts of the island, sites that would have been better used for homesites. A number of the lower-elevation sites used for houses belonging to the club were threatened by a trend of severe shoreline erosion that began in 1982. Hugo proved the importance of setback when houses far back from the beach suffered less wave and water damage.

Hurricane Hugo completely submerged Isle of Palms, and its dunes were eroded down to planar level (appendix D, ref. 32). Downbursts of wind caused extensive damage in the interior of the island. Beachfront houses were destroyed, and the shore adjacent to Breach Inlet was eroded. The published erosion rate in the vicinity of Breach Inlet is 2.6 feet per year. Obviously, beachfront areas should be avoided. Shoreline engineering structures built prior to Hugo, such as the groin field to the north and

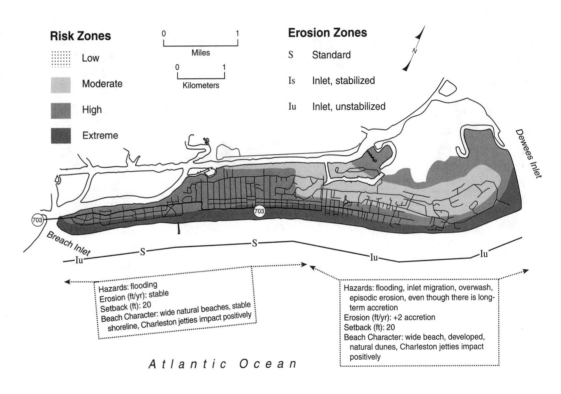

Risk Zones

Low

Moderate

High

Extreme

Erosion Zones

S   Standard

Is   Inlet, stabilized

Iu   Inlet, unstabilized

Hazards: flooding
Erosion (ft/yr): stable
Setback (ft): 20
Beach Character: wide natural beaches, stable shoreline, Charleston jetties impact positively

Hazards: flooding, inlet migration, overwash, episodic erosion, even though there is long-term accretion
Erosion (ft/yr): +2 accretion
Setback (ft): 20
Beach Character: wide beach, developed, natural dunes, Charleston jetties impact positively

Atlantic Ocean

RM 11. Isle of Palms.

6.18. Left: Low sand dunes on Isle of Palms shoreline. The sand fence in the foreground has been successful at sand accumulation.

6.19. Right: View of Sullivans Island beyond Breach Inlet. In the background is the entrance to Charleston Harbor; beyond that lies Morris Island.

the seawalls, indicate areas of past erosion and predict future beach loss.

Most areas of this island are backed by low sand dunes (fig. 6.18). Beach width increases toward the southern end of the island. So much new land accreted on the south end of the island that the original developer returned to lay claim to the land that formed between the once beachfront cottages and the new,

more distant shoreline. An important lesson here is that people buying beachfront property should always check their deeds to see who will have title to land that might accrete along their lot front, and who has title to the lot when it is under water.

The pre-Hugo development of the inner part of Isle of Palms was unique in that roads actually bent around ancient live oaks, preserving the protective forest, and were built on top of, rather than through, old stabilized dunes, maintaining the protective ridge topography. Much of the central and seaward axis of the island is above 10 feet in elevation, which was not sufficient to prevent flooding during Hugo. The new bridge provides a better hurricane

evacuation route, eliminating the former potential traffic bottleneck at the Breach Inlet Bridge (fig. 6.19) when the only route off the island was via Sullivans Island. Pre-Hugo storm surges had generated water levels 8 to 11 feet above sea level that flooded many parts of the island, so Hugo's storm surge should have been no surprise to those in the development on the back side adjacent to the waterway and along parts of the old escape route. Much of the island falls into the extreme- to high-risk categories, and the rebuilding that followed Hugo makes it probable that future losses will be greater rather than lower.

Sullivans Island (RM 12)

This 4.3-mile-long island is a suburb of
Charleston (fig. 6.19) that is zoned to maintain
its single-family housing character. Fort
Moultrie and associated fortifications on the
south end of the island overlooking the entrance to the harbor were maintained from the
time of the American Revolution through
World War II. The fort is the site of a famous
Revolutionary War battle in which the Americans used palmetto logs for fortification instead of oak. The British cannonballs sank into
the soft palmetto logs rather than shattering
the fortification, while the American cannon
fire tore into the British ships. In commemora-
tion of this victory South Carolina is called the
Palmetto State.

   The fort's long occupancy resulted in a good
record of the frequency and impact of hurricanes on this island. Destruction and loss of
life on Sullivans has occurred during several
hurricanes. Hugo completely submerged the island, eroded dunes, and carried overwash into
the interior. The damage pattern was
islandwide, as it was on Isle of Palms, but the
greatest destruction was in the shore zone. Earlier hurricanes caused similar destruction. In
1833 the entire island was inundated by 4 to 5
feet of water.

   Sullivans Island is also much like Isle of
Palms in terms of risk classification. The area
adjacent to Breach Inlet is subject to erosion
problems as well as floods. Attempts to stabilize the inlet have not been particularly suc-

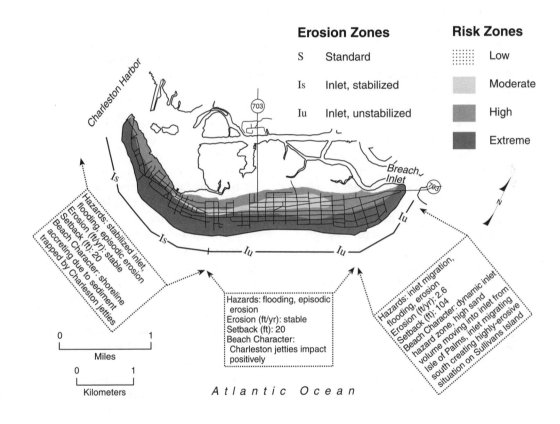

Erosion Zones

S   Standard

Is   Inlet, stabilized

Iu   Inlet, unstabilized

Risk Zones

Low

Moderate

High

Extreme

Hazards: stabilized inlet,
 flooding, episodic erosion
Erosion (ft/yr): stable
Setback (ft): 20
Beach Character: shoreline
 accreting due to sediment
 trapped by Charleston jetties

Hazards: flooding, episodic
 erosion
Erosion (ft/yr): stable
Setback (ft): 20
Beach Character:
 Charleston jetties impact
 positively

Hazards: inlet migration,
 flooding, erosion
Erosion (ft/yr): 2.6
Setback (ft): 104
Beach Character: dynamic inlet,
 high sand
 hazard zone,
 volume moving into inlet from
 Isle of Palms, inlet migrating
 south creating highly-erosive
 situation on Sullivans Island

0   1
Miles

0   1
Kilometers

Atlantic Ocean

RM 12. Sullivans Island.

cessful. The north-to-south longshore drift causes the inlet to migrate into Sullivans Island. Property immediately adjacent to the inlet is in an extreme-risk zone, and all of the area fronted by the groin field should be looked on with caution. Dune lines on this island tend to be irregular, and gaps or low dunes should be avoided. The low dunes were too small to be effective against Hugo.

The entire island lies within the 100-year flood zone, which places all of the island in either the extreme-risk or the high-risk zone. The V-zone on the Federal Emergency Management Agency's (FEMA) Flood Insurance Rate Maps (FIRMs) extends nearly halfway through the island. For the most part, significant forest cover is lacking, again placing most sites in the high-risk category.

The south-central part of the island has a long history of accretion, but sand trapping and accumulation accelerated after the north Charleston Harbor jetty blocked north-to-south longshore sand transport. The beach is wide and backed by irregular low dunes. Sand has accreted along the shore inside the jetty. The jetty also has created offshore shoals in front of Sullivans Island.

The nearly total development of Sullivans Island, the storm-surge flood threat, and the fact that storm-tide levels of hurricanes in 1893, 1940, 1959, and 1989 were all equal to or higher than the low elevations along the present escape route indicate that hurricane warnings should be taken very seriously. The

new bridge eliminates the problem that occurred during Hugo when the old Ben Sawyer swivel bridge (fig. 3.4) was severely damaged and stuck in the open position, but its presence should not give residents a false sense of security. Early evacuation is the rule.

Morris Island (RM13)
We described the extreme instability and rapid migration of this island in chapter 1. Morris Island today is a narrow strip of land 4 to 5 miles long, consisting only of washover terraces and a dredge spoil impoundment. It is bordered on the north by Charleston Harbor and on the south by Lighthouse Inlet. The most significant event in Morris Island's history was the fierce battle and three-month siege to take Fort Wagner during the War Between the States; it was the greatest battle in South Carolina's history. Today, Morris Island is literally world famous not for what it is or was, but simply for its physical transformation. In geological terms, Morris Island moved in the blink of an eye.

In fact, the events on Morris Island illustrate just how rapidly risk vulnerability can change (fig. 1.3). Morris was once a large beach-ridge island, substantially wider than Folly Island; its ridge complex originally included Long Island and part of Folly Island, where the old Coast Guard Station now stands. The beach ridges formed about 4,000 to 5,000 years ago when the rapid sea level rise slowed, allowing the ridges to accumulate and become the land-

ward shield of the huge Charleston Harbor delta.

Anyone looking at Morris Island today finds it difficult to imagine the dune-ridged island that was a highly coveted stronghold during the Civil War. The Confederates held historic Fort Wagner to protect the supply lines into Charleston Harbor, but they were unable to withstand the three-month siege by the Union's army and navy attacking from adjacent Folly Island. Interestingly, the initial attack would have succeeded and there would have been no siege if it had not been for the tide! At low tide the Union soldiers waded and were ferried across Lighthouse Inlet to attack by land. The fort was about to fall when the tide began to come in, leaving the soldiers stranded within rifle-shot distance of Fort Wagner but without the support of the troops on Folly Island. A bloody three-month battle ensued, and the Union Army eventually won.

After the Civil War, the construction of the Charleston Harbor jetties, completed in 1896, spelled doom for Morris Island. The jetties blocked sediment transport to Morris Island, cutting off the natural nourishment and causing extreme erosion. The south Charleston jetty, in fact, is anchored to northern Morris Island. So, like a great windshield wiper, the southern shoreface has swung landward, three quarters of a mile in 50 years. In recent times a spoil impoundment dike was constructed that slowed the landward migration of Morris.

Today, none of the great beach ridges remain. The Morris Island lighthouse (fig. 1.4), built between 1874 and 1876 in the marsh behind the island, now stands 2,000 feet offshore, watchful testimony of one of the good works of man gone bad. Modern erosion rates are approximately 20 feet per year for the area south of the jetty, and 6 feet per year at the jetty. The jetties have caused slow but steady accretion to both Sullivans and Morris Islands inside the harbor. Privately owned Cummings Point, which extends north from the south jetty, has exhibited shoreline fluctuations of several hundred feet over the last 50 years. The high erosion rate, very low elevation, lack of significant vegetation, and frequency of overwash make the island totally unsuitable for development; all of the area is at extreme risk.

Folly Island/Folly Beach (RM 13)

Folly Beach is billed as "Charleston's Original Beach Town," or more recently, Charleston's playground, but from the perspective of coastal hazards a better title might be "The Town of Tomorrow's Coastal Problems Today." Like Morris Island, Folly is a dramatic example of what can happen when humans interact with the coastline. Its ongoing saga is of a town engaged in a battle with nature which has its roots in history, its drama in property loss, and its moral yet to be learned. Erosion was common on Folly well before the construction of the Charleston Harbor jetties, as

RM 13. Folly Beach

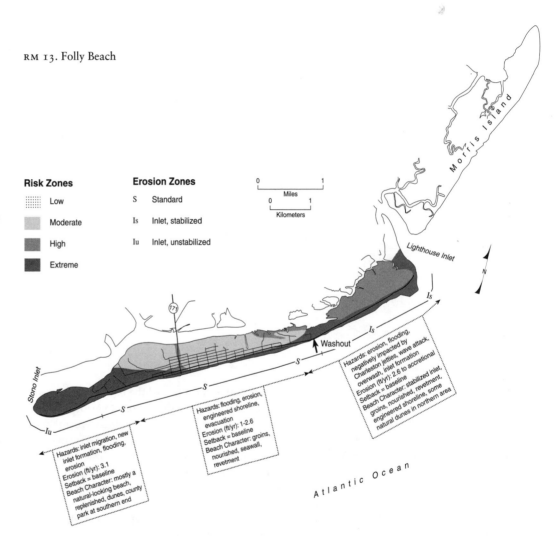

Risk Zones

Low

Moderate

High

Extreme

Erosion Zones

S   Standard

Is   Inlet, stabilized

Iu   Inlet, unstabilized

0 ——— 1
Miles

0 ——— 1
Kilometers

Hazards: inlet migration, new inlet formation, flooding, erosion
Erosion (ft/yr): 3.1
Setback = baseline
Beach Character: mostly a natural-looking beach, replenished, dunes, county park at southern end

Hazards: flooding, erosion, engineered shoreline, evacuation
Erosion (ft/yr): 1-2.6
Setback = baseline
Beach Character: groins, nourished, seawall, revetment

Hazards: erosion, flooding, negatively impacted by Charleston jetties, wave attack, overwash, inlet formation
Erosion (ft/yr): 2.6 to accretional
Setback = baseline
Beach Character: stabilized inlet, groins, nourished, revetment, engineered shoreline, some natural dunes in northern area

Washout

Lighthouse Inlet

Stono Inlet

Morris Island

Atlantic Ocean

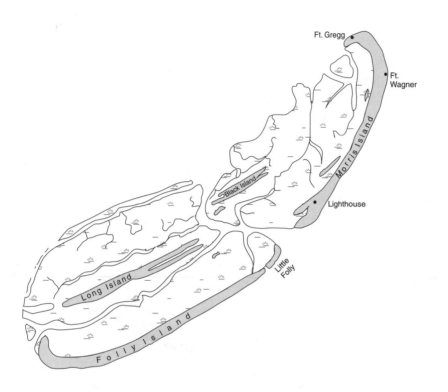

Ft. Gregg

Ft. Wagner

Morris Island

Black Island

Lighthouse

Long Island

Little Folly

Folly Island

6.20. Civil War–era map of Folly Island, Little Folly, and Morris Island. Though replaced by maps of better quality and detail due to the Union occupation, this map is more accurate than it first appears. Inlets have shifted greatly and Morris Island has migrated three quarters of a mile, primarily in recent years. *Source:* Modified from *Time and Tide on Folly Beach, South Carolina (a History)* (appendix D, ref. 133).

lands. The northern third, beyond the flexure known as "the washout," is a low-lying strip of land that formed 3,000 to 4,000 years ago. At one time, Little Folly, as this northern segment is called on old maps, could only be reached at low tide by crossing the washout (itself an old inlet). This remained the case until the 1950s, when a road was finally built on an artificial land bridge. Interestingly, nature has not yet accepted the fact that this zone is no longer an inlet. Three times in the late 1980s storms reopened the washout, and Hugo washed out eight houses as well. It's called "washout" for a reason!

The part of Folly Island from the washout downcoast to the county park was formerly known as Big Folly Island. This island formed by the slow accumulation of sand in dune ridges about 4,000 to 5,000 years ago when the rate of sea level rise slowed. Its beach ridges are relatively high in elevation and in their natural state are covered with pine forest. In fact, the Old English name *folly* or *volly* means "a tree-crested dune ridge," a reference

suggested in an 1848 report (appendix D, ref. 65). The trend greatly accelerated as a result of the jetties, however, both because the longshore sand supply was blocked and because the protective offshore bar-shoal complex disappeared (fig. 1.2). Although settlement dates to earlier times, the present resort community development began in 1920. From that time onward, oceanfront lots and buildings have been calling attention to the long-term processes of retreat; the so-called erosion

problem was born, and the battle was on. A half-century's worth of groins, various kinds of walls, revetments, and rubble were either destroyed and strewn about by Hurricane Hugo in 1989 or buried by the beach nourishment of 1993. By 1995, some of this old debris was reappearing as the beach was again quickly disappearing (fig. 4.2).

Folly Island is separated from Charleston Harbor by the remnant of Morris Island and Lighthouse Inlet and is actually two distinct is-to the first sight of land a sailor would see when approaching Charleston Harbor. There are numerous islands in the Carolinas originally called "folly."

During the siege of Charleston by Union forces, Folly Island was the temporary home to more than 20,000 Yankee soldiers (fig. 6.20). Many of today's drainage ditches, causeways, revetments, batteries, and even Folly's roads are the engineering works of the Union Army. A fierce battle raged between the Union artillery at the old Coast Guard Station and the Confederate batteries across Lighthouse Inlet. Hurricane Hugo's erosion uncovered this forgotten battlefield.

Folly's erosion problems began before the town even existed. The construction of the Charleston Harbor jetties in the 1890s was driven by the need to improve the harbor; at the time, there were no rows of beachfront houses downdrift. The 3-mile-long rock jetties block the southerly drift of beach sand by longshore currents. Sediment accumulates on the updrift side of the north jetty and spills over into the channel. Instead of being pumped over the south jetty, the sand is dredged and taken several miles offshore to be dumped. Thus, material that would naturally replenish Morris and Folly Islands is lost from the system. As noted above, however, the jetties are not the sole cause of Folly's erosion. Shoreline-change maps of the area (appendix D, ref. 70) show erosion rates of about 10 feet per year in the 1800s for the main part of the island, and

up to 50 feet per year for the area near Stono Inlet.

For a time, the jetty construction actually caused Folly's beach to accrete. The jetties caused the collapse of an immense offshore tidal delta which extended from Breach Inlet to the washout. Much of this freed sand came ashore, resulting in a 30-year period of accretion on Morris and Folly Islands. Old maps show plans for a subdivision and four new roads seaward of Ashley Avenue! The tidal delta was a finite sand source, however, and the sand ran out before the new community could be built. Following several hurricanes in the 1930s and 1940s, residents had a new view to the northeast. Storm waves, which had always broken on the delta shoals, now broke right on the beach. Folly's erosion problem was back.

House Document 156, prepared by General Douglas MacArthur and presented to the 74th U.S. Congress in 1935, should be required reading for anyone buying property on Folly Beach. The report, titled "Beach Erosion at Folly Beach, S.C.," documents shoreline changes and erosion rates over more than 84 years. A map record demonstrates average erosion rates of 7 feet per year for most of the island, and rates as high as 51 feet per year adjacent to Stono Inlet. The report overlooks the long-term significance of these figures, perhaps because erosion had temporarily slowed as sand from the collapsing tidal delta continued to wash ashore. At the time there was still a

wide area between the eroding beach and the island's houses. But short-term trends are misleading.

Storms in 1933 and 1934 caused serious beach erosion and recession, threatening what was then the second row of houses, destined to become the front row. The 1940 hurricane eroded 75 feet of real estate. After World War II the South Carolina Highway Department saw the installation of groin fields as the way to beat the sea, and more than 40 groins were constructed between 1949 and 1970. In spite of the groins, however, 1959 saw another 35 to 50 feet of land lost, along with most of the remaining houses seaward of Arctic Avenue. Hurricane David in 1979 did little damage elsewhere in South Carolina, but the damage on Folly was severe. After the storm, only two houses remained on the seaward side of Arctic Avenue. These houses were combined after David to form the Atlantic House, where one could dine as the waves rolled under the building.

Hurricane Hugo exceeded David's work in building removal. The 1989 storm wreaked havoc on Folly Island. In addition to the destruction at the washout, beachfront houses were destroyed or heavily damaged, along with their so-called protective structures. The Atlantic House Restaurant was destroyed. Revetments were flattened and dunes were leveled as the surge overwashed and flooded the island, damaging houses in the interior. Shore-perpendicular streets became avenues for overwash, and a layer of sand was carried 250 feet inland. The water level was more than 12 feet above sea level (appendix D, ref. 32), and the beaches were pummeled by 20-foot waves. Although it was to the left of the storm's eye, Folly suffered as much erosion as beaches to the right of the eye. The Holiday Inn, "protected" by a massive concrete seawall, was flooded and sustained major first-floor damage. Both overwash and ebb-flow currents eroded streets, destroyed dunes, and undercut foundations.

The collective result of storm destruction and ongoing erosion is that Arctic Avenue now ends before 13th Street East, although it formerly ran to 16th Street East. The old tax maps tell the story: Atlantic Avenue lost a row of lots now beneath the sea, part of Arctic Avenue amputated, and half or more of a second row of lots removed. The losses are not just squares on a map; they were real houses and homes, real dollar investments, and real dreams lost. It's costly to battle nature.

Storm waves have not dampened the enthusiasm for seaside living. Folly Beach may lose frontally, but it has grown laterally and continues to grow in depth. The last remaining stands of forest are being developed. The permanent population is now more than 1,700, and a summer weekend may see 3,000 people on the island. Voting residents struggle to keep the social flavor of the island through strict zoning laws that permit only single-family dwellings throughout most of the island and limit the height of structures (the Holiday Inn excluded), and by battling developments such as the prefabricated condominiums once slated to be built on the island's high-risk southern spit. Ultimately, the spit became a county park that provides much-needed access and public recreation for the thousands of visitors. Ends of city streets provide more restricted public access to the beach. And the erosion continues.

After Hurricane David the headlines claimed $1 million in damages at Folly Beach. David demolished many of the private makeshift protective structures fronting beach buildings. More recent northeasters removed additional sand from beneath beachfront houses. By the early 1980s the magnitude of the problem was reflected in the hodgepodge of private revetments, walls, sand grabbers, and a kitchen sink that had been added to the state-maintained groin system. A 1979 study by the Army Corps of Engineers (appendix D, ref. 135) projected that in 50 years the shoreline would be well into the island, behind these weak defenses—and the houses and lots and roads they were meant to protect. Old tax maps and previous shoreline history support such projections.

And the seawall saga has not ended. Folly Beach enjoys a unique privilege under the Beachfront Management Act because it is exempted from two important sections of the act: the retreat policy and the seawall prohibition. Most of Folly is armored, and, under the exemption, seawalls on the island may be rebuilt if they are destroyed. Folly planners hope that the present shoreline can be maintained by beach nourishment. Should this prove not to be possible or feasible, other sections of the law require that Folly rejoin the state's retreat plan. If Folly loses its exemption it will eventually lose its seawalls. Then the Corp's 1979 projection of the fifty-year shoreline will become a reality.

The northern portion of Folly Beach, from Lighthouse Inlet to about halfway from the washout to the inlet, is a Stabilized Inlet zone (see chapter 10 for an explanation of the erosion zones established by the Beachfront Management Act). The extensive engineering works at the old Coast Guard Station anchor the north end and, along with the revetted and nourished washout (fig. 6.21), hold the shoreline in place to keep it from following Morris Island and marching toward the mainland. The official erosion rate for this area is 2.6 feet per year. Even from the most optimistic point of view, the northern part of Folly Island to just south of the washout is an extreme-hazard area.

Except for the county park, the rest of Folly, including the washout, is in a standard erosion zone. Erosion rates range from 2.6 feet per year in the north, through 1.9 feet per year in the center, to 1.0 feet per year at the south end. The county park at the southwest end is an unstabilized inlet zone with 3.1 feet per year of erosion. Remember that although Folly is highly erosional, it is stabilized by seawalls that may be maintained in their current loca-

6.21. Left: The Folly Beach washout is marked by the rock revetment and the grass-planted artificial sand fill. Note how narrow (nonexistent, really) the island is here. The revetment was rebuilt with much larger stone after Hugo and now "protects" the road. Dune grass and sand fencing were emplaced after the 1993 beach replenishment project. The houses in the background occupy the former Little Folly Island (see fig. 6.20) and face not only multiple coastal hazards but the certainty that their access over the washout will be washed out again.

6.22. Right: The site of this quaint old Folly Beach cottage was chosen wisely; the high, forested dune ridge gives natural protection from floodwaters.

tion. Therefore, these erosion rates may be on the low side.

The main part of the town is at moderate risk atop the higher ridges, at high risk in the lower A-zone, and at extreme risk on beachfront and V-zone property. Numerous old cottages in the middle and on the back side of the island that sit on high ground and under a protective canopy of vegetation attest to the inner island's relative stability (fig. 6.22). New development in the forested portion of the south end of the island will be at mostly moderate to high risk, and each site requires evaluation. The spit is an extreme-risk zone. Stono Inlet has oscillated through time, and the spit

could be cut off by new inlet formation during a hurricane.

There is public beach access along most of Folly Island, provided by 28 accessways located at street ends and at mid-block points. In addition, there are five public parks on Folly. The newest is a city park at the north side of Center Street along Folly River. The old Coast Guard Station, recently acquired by the Charleston County Parks and Recreation Commission, is the fourth county park on Folly. It is slated to be developed only as a nature preserve. The first county park was Folly Beach County Park, adjacent to the Stono Inlet. Folly's new pier (fig. 6.23) is also a county

6.23. Low-tide photo of the pier and Holiday Inn seawall at Folly Beach. Compare the narrow beach in front of the seawall with the wide beach in the area without a wall. At high tide there is no beach in front of the structure. In 1993 replenishment sand was level with the top of the wall. By 1995 the beach was about 15 feet below the top of the wall!

park, and a former campground by the boat landing to the south of Center Street is a county park with boat access and related facilities. Folly has been called "Charleston's Playground," and thanks to the Charleston County Parks Commission, it provides the finest beach and water access of any area of its size in South Carolina.

Bird Key (RM 13)

Living on a barrier island involves not just enjoying the amenities that come from occupying a costly parcel of land but also protecting those amenities. This can have repercussions far beyond the beachfront. Bird Key is an ephemeral feature of the large ebb-tidal delta in the mouth of Stono Inlet. The shoal was the largest brown pelican rookery in the southeast until it virtually disappeared. The 1993 nourishment of Folly Beach put a large amount of sand into the southerly drift system of the island. As the Folly spit grew into Stono Inlet, the channel was displaced, and its flow then eroded Bird Key. Although it had verbally assured the local community that Bird Key would not be affected by the nourishment

project, the Corps of Engineers made no effort to maintain or restore the rookery. Pelicans may be regarded as a flying amenity of barrier islands or as big, ugly birds that are expendable. The latter view may come back to haunt all the species that inhabit barrier islands.

Oak Island and Long Island (RM 13)

In the extensive marsh behind Folly Island lie old beach or dune ridges of either former barrier islands or ancient marsh islands. These islands stand above the marsh and are well forested by oak and associated vegetation. They are at lower risk than oceanfront, but they are in the flood zone and are vulnerable to hurricane-force winds. Other problems their residents face include groundwater supply, waste disposal, and congested evacuation routes. In spite of their vulnerability, the increasing demand for building sites next to the water has brought development to these back islands. Consider them high-risk areas with respect to potential damage from coastal hazards.

Kiawah Island (RM 14)

Kiawah Island is a 10-mile-long beach-ridge barrier island located about 30 miles south of Charleston. People have occupied this island since before whites and blacks came to the continent, and Kiawah takes its name from the Indian tribe that once lived here. Kiawah and Seabrook Islands together form the oceanfront barrier seaward of Johns Island (formerly St. John's Island) and Wadmalaw Island. Johns and Wadmalaw Islands are rural areas within a maze of winding streams and broad marshes. The migrating Captain Sams Creek and a marshy no-building zone separate Kiawah from Seabrook. Kiawah's northern boundary is Stono Inlet. The island comprises numerous parallel rows of massive dunes, the most landward of which formed during the Pleistocene. The outer dune ridges and seaward portion of Kiawah Island are younger.

Undeveloped Kiawah was purchased in the mid-1970s by the Kuwait Investment Company, which commissioned two predevelopment studies: an assessment of the island's environmental resources and a development plan. Special ecosystems were to be left undisturbed, vast tracts were to remain undeveloped, and no houses were to be built in the dune field seaward of the maritime forest. Kiawah was to be a showcase of wise and sensitive development of a pristine barrier island. That was two owners and one war ago.

Environmentally, how does today's developed Kiawah Island rate? Still very good. The island has been subdivided considerably beyond the original plan, but development density remains low. On much of the island even a view of the house next door is obscured by relatively undisturbed forests and dunes (fig. 6.24). Covenants still prevent building anywhere near the primary dunes. Most important, development on the northeastern half of the island has been restricted to a golf course and its associated amenities. There are few structures in this low-lying area to be threatened by shoal attachment, inlet migration, and storm flooding.

A great deal of information is available for evaluating risk on this island (appendix D, ref. 136), and prospective buyers should not hesitate to ask sales representatives to see the reports (reputable companies will have copies available). Kiawah's development remains low density, and ecologically sensitive areas and areas in danger of erosion will not be developed. The latter includes the northeast end of the island, shown in RM 14 as an extreme-risk zone occupied by Stono Inlet and ancestral inlets since 1661. Although the trend since then has been accretion, beach ridges have been truncated by erosion, and marsh clays are exposed on the beach. Erosion rates adjacent to Stono Inlet have been measured at 55 feet per year, although the state's published rate to determine setback is around 5 feet per year (locally up to 12 feet per year). Dunes are absent along the beachfront, and overwash is frequent. Two small tidal inlets occasionally open up in this area. The adjacent coast to the southwest in front of the line of cat's-eye ponds is an area of extreme risk because of potential flooding, reactivation of old inlets, and overwash. Inland in this area, the forested dune ridges are commonly 10 to 15 feet in elevation and at moderate to high risk.

The central portion of Kiawah is slightly accretional (less than 1 foot/year) and generally stable in the long run. The accretion is from sand derived from erosion at the north end of the island, shoals in the ebb-tidal delta that periodically detach and bring sand onshore, and probably from the high erosion of Folly Island over a longer time span. The nearshore zone is characterized by protective dunes but is still high risk, whereas in the central part of the island, along the higher forested ridges, the risk for property damage is moderate to low. Keep in mind that even though the beach in front of the main body of the island may be accreting and buildings are required to be set back behind a well-developed line of dunes, the frontal developed area is within the 100-year flood zone and at elevations generally less than 10 feet. The low eleva-

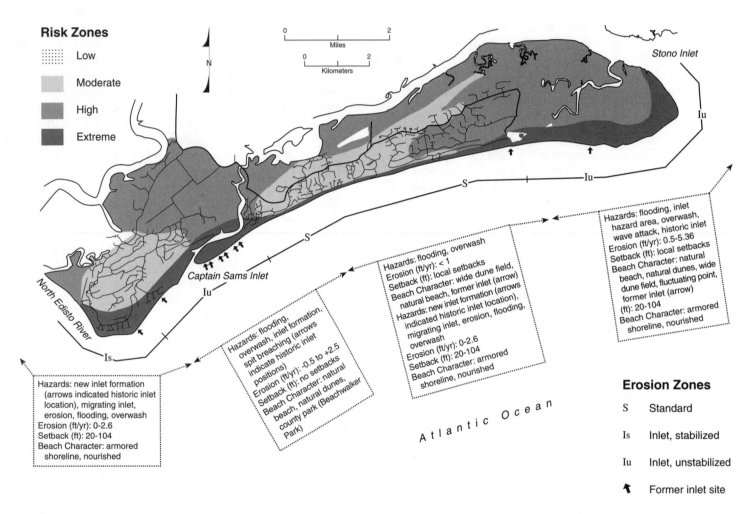

Risk Zones

Low

Moderate

High

Extreme

0   2
Miles

0   2
Kilometers

N

Stono Inlet

Iu

S

Iu

Captain Sams Inlet

Iu

North Edisto River

Is

Atlantic Ocean

Hazards: flooding, inlet hazard area, overwash, wave attack, historic inlet
Erosion (ft/yr): 0.5-5.36
Setback (ft): local setbacks
Beach Character: natural beach, natural dunes, wide dune field, fluctuating point, former inlet (arrow) (ft): 20-104
Beach Character: armored shoreline, nourished

Hazards: flooding, overwash
Erosion (ft/yr): < 1
Setback (ft): local setbacks
Beach Character: wide dune field, former inlet (arrow) natural beach, former inlet (arrow)
Hazards: new inlet formation, indicated historic inlet location, migrating inlet, erosion, flooding, overwash
Erosion (ft/yr): 0-2.6
Setback (ft): 20-104
Beach Character: armored shoreline, nourished

Hazards: flooding, overwash, inlet formation, spit breaching (arrows indicate historic inlet positions)
Erosion (ft/yr): -0.5 to +2.5
Setback (ft): no setbacks
Beach Character: natural beach, natural dunes, county park (Beachwalker Park)

Hazards: new inlet formation (arrows indicated historic inlet location), migrating inlet, erosion, flooding, overwash
Erosion (ft/yr): 0-2.6
Setback (ft): 20-104
Beach Character: armored shoreline, nourished

Erosion Zones

S   Standard

Is   Inlet, stabilized

Iu   Inlet, unstabilized

↖   Former inlet site

RM 14. Kiawah Island and Seabrook Island

6.24. Interior homesite on Kiawah Island. Houses built on the forested interior dune ridges of Kiawah have the dual advantage of privacy and protection from natural hazards.

tion is partly the result of leveling done for rice and indigo cultivation in the plantation days.

Kiawah was the site of a dramatic inlet stabilization project in the 1980s. South Carolina's inlets range in size from small swashes in the north to the gigantic sounds of the low country. Each size of inlet exhibits a different manner of bypassing, or transporting sand from the updrift side to the downdrift side. Medium-sized inlets bypass sediment that accumulates on the downdrift end of the upper island. Since the inlet's size must remain constant, growth of the spit forces the inlet to migrate downcoast. This inlet migration continues in a cycle, usually over decades, until the ebb flow of an extremely high tide, such as a storm surge, sweeps over the spit and carves a new channel. The part of the spit that is cut off accumulates on and nourishes the downdrift island, and the process begins again. Obviously the inlet migration zone is a hazardous place for any type of construction and should be considered the "inlet hazard zone" for management purposes.

In the early 1980s an innovative form of coastal engineering was used to stabilize the shoreline between Kiawah and Seabrook Islands. The progressive downcoast migration of Captain Sams Creek was causing abnormally high erosion on Seabrook Island, undermining numerous properties and much of a golf course. Studies revealed that the historic migration cycle of Captain Sams Inlet was on the order of 80 years. The bad news was that several decades remained before the expected end of the cycle (not to mention that the cycle itself depended on something as variable and unpredictable as a hurricane). The Coastal Science and Engineering Division of the Research Planning Institute of Columbia, South Carolina, proposed and engineered a solution (appendix D, ref. 137). The natural inlet was closed, and an artificial new inlet was created. This had never been tried before, and there was some risk involved. So far, however, the project has been a marvelous success. The huge quantity of sand released from Kiawah's spit has moved downcoast as predicted to heal the scars of the previous erosion episode and has added hundreds of feet of land seaward of the seawalls along the former inlet location (fig. 6.25). The project was performed again by CSE-BAIRD in the spring of 1996. It serves as a great example of success for soft shoreline engineering. Beach Walker County Park is located on Kiawah's spit and is not affected by the inlet relocation.

Kiawah awaits two tests. The first will be a direct hit by a category 3 hurricane (all bets are off in a category 4 or 5 storm). The second will be to see if the present and future developers stick to the "plan" or turn the entire 3,300 acres of high ground into the "scenery of sameness" of other developed islands. Good work so far, Kiawah; and good luck in the future.

Seabrook Island (RM 14)
If Kiawah Island is an example of moderate-density development that avoids risky inlets and unstable beaches, then Seabrook Island is its antithesis. Development has been placed in unstable areas, and problems are the result.

Seabrook Island, a barrier island roughly 4 miles long, is adjacent to North Edisto Inlet and is separated from Kiawah Island to the north by Captain Sams Inlet. Seabrook was incorporated in the late 1980s, but it remains a private development of mixed single-family and multifamily structures. A single massive rock revetment begins at Pelican Watch Villas on the inlet, goes around the clubhouse, and

extends past Renken Point. This area is classified as a stabilized inlet zone.

Development of Seabrook began in the 1970s after two studies, both flawed, found the island accretional. The flaw lay in the time scale the investigators considered. Experience has shown that erosion and accretion trends are cyclical, and the larger the size of the inlets adjacent to an island, the longer the cycle. On

6.25. Landward view of the former margin of Captain Sams Inlet separating Seabrook Island and Kiawah Island. Anticipating nature, the new inlet was relocated to the north, releasing sand that then welded on to the beach at Seabrook. The house here once had a sea view; now it has a marsh view.

Seabrook, the erosional cycle actually began after planning and development had commenced! The northern channel of the North Edisto Inlet has shifted landward (north) by several hundred feet (appendix D, ref. 137). The large revetment wall extending for a mile around the clubhouse was meant to stabilize the inlet's north shore. Although the wall is being flanked along the inlet, no buildings are threatened yet. In 1990 a renourishment project placed 685,000 cubic yards of sand on the beach. Residents hoped this would encourage the marginal channel to shift seaward. Apparently this shift did not occur, and much of the new sand has migrated out of the area. The revetment is exposed from Renken Point to the North Edisto Inlet. Published erosion rates in this reach are as high as 2.6 feet per year.

Seabrook Island faces a different problem to the northeast. As the spit at the southwest end of Kiawah Island grows, it forces Captain Sams Inlet to the southwest. This migration erodes Seabrook Island at a rate of 250 feet per year (appendix D, ref. 131). In 1983 the inlet was artificially relocated upcoast by several thousand feet (see above, under "Kiawah Island"), relieving some of the erosional stress. The north beach area (north of Renken Point) has accreted into one of the widest beaches in the state. Houses armored with seawalls against the migrating inlet now sit nearly 800 feet from the surf (fig. 6.24).

A permit to relocate Captain Sams Inlet once again was approved in January 1996. If all goes according to schedule, the project will be completed early in 1996. There is still much sand in the north beach area, but the inlet has migrated well into Seabrook again. The thirteen-year renourishment interval is probably the best such record in the state.

To its credit, the town of Seabrook Island has used local funds for its beach preservation projects. The area fronting the North Edisto Inlet appears stable for the present. How long the revetment holds is up to the weather gods. Should it be necessary, there is still ample room to relocate the clubhouse and facilities landward. In view of the rising sea level, we should not look for the return of accretion anytime soon.

Deveaux Bank, the large, low-lying sandbar in the mouth of the North Edisto Inlet, blocks some wave energy and shelters part of the Seabrook Island shore. The bank, however, comes and goes. One study in the middle 1970s recorded more than 20 feet of loss on a single high tide. It was removed completely by Hurricane David in 1979 but has since recovered and developed stabilizing vegetation.

Buildings constructed adjacent to the beach in the early development phase were soon threatened, and the rock revetment began to grow. Almost at its inception the Seabrook Island development was wed to shoreline engineering. Seawalls and sand added to the beach have not solved the problem, and the pressure to outengineer nature is extending beyond the island. Proposals include the artificial nourish-

ment of Deveaux Bank and the closing of Captain Sams Inlet. Low-risk sites exist inland, well away from the beach. The immediate coastal zone is high risk (RM 14).

Deveaux Bank (RM 15)
As noted above, Deveaux Bank is a sandbar that owes its existence to the inlet. It is not a barrier island. Rapid accretion in recent years has increased the bank's size to the point that green vegetation is now visible even at high tide. It has been suggested that the growth of the shoal is linked to the landward migration of North Edisto's ebb channel, and thus to the erosional stress on Seabrook Island.

Development here always has been out of the question. The bank has been an important brown pelican rookery and, when it accommodates nesting birds, is managed as a wildlife refuge by the National Audubon Society. In addition, the bank functions as a sand reservoir.

The Low Country

The southern one-third to one-half of South Carolina's coast is commonly referred to as the low country. As the term implies, this region is only slightly above sea level. The land is gently subsiding, or lowering its elevation. The vast amount of sediment that flows into the area keeps pace with the subsidence, but just barely.

The tide range in the low country is mesotidal—greater than 2 meters. Although not unheard of elsewhere, such a large tidal range is somewhat uncommon in the United States. It is caused by the concave shoreline configuration known as the Southeast Georgia Embayment, which descends southwestward from Cape Hatteras, North Carolina, through the Carolinas, but turns to the southeast in the Savannah River area. The concave shape focuses and magnifies the tidal extremes within the embayment.

Mesotidal coastlines are characterized by barrier islands that are relatively shorter and "stubbier" than barrier islands in microtidal areas. The islands are separated by large inlets and front immense marshlands. Because they are not fed by major rivers, Port Royal Sound and St. Helena Sound have been called the largest tidal inlets in the world.

It is not clear where the subsidence of the low country begins, though it certainly includes all of the Charleston area. We refer to it here because it includes the Edisto Rivers, which drain the low country and flow into the Atlantic Ocean.

Charleston County, continued

Botany Bay Island (RM 15)
This small island lies across the North Edisto River Inlet from Seabrook Island. At one time, its development seemed imminent. A water desalinization system was constructed but was never used because the development plans were abandoned. The fact that development was even considered here illustrates the serious state of affairs along the South Carolina coast. The National Audubon Society advocates preserving this fragile island.

Few islands could serve as a better example of where not to build. Botany Bay is woefully small and has only a mile of beach. It lies at the very mouth of a major inlet, an area where the instability of sediment bodies is known. Deveaux Bank, the shoal to the east, might offer some protection from open-ocean waves, but the bank comes and goes, and the wave refraction patterns constantly change.

Erosion is known to be severe, averaging tens of feet per year. More than 400 feet of beachfront was lost between 1949 and 1973, mostly along the island's eastern curve (appendix D, ref. 47), and the island may have lost as much as 40 percent of its total area since 1939. Hurricane David in 1979 is reported to have eroded as much as 50 feet.

Unlike most other barrier islands, the sand ridges on Botany Bay Island run perpendicular to the coast rather than parallel. As a result, the ridges offer little protection from storm winds and flooding. Most of the island is less than 10 feet above sea level; that is, it lies in the flood zone and would be almost totally submerged in less than a 100-year flood. The ridge orientation also allows frequent overwash.

Changes in the tidal drainage around the island are also common. In fact, the island owes its separate identity to the formation of a new inlet cut during a 1948 storm. South Creek, which formerly ran behind this island into the North Edisto River, was intersected by a breach through the island front and marsh, cutting the island off from Botany Island (Botany Plantation) and the contiguous marsh of Edisto Island. This larger island was also called Botany Bay Island, which may generate some confusion when comparing older and new maps. The area around the 1948 inlet changes frequently.

Botany Island (Botany Plantation; RM 15)
Botany Island is privately owned and undeveloped. Its short beachfront is located between two inlets, and its very low elevation above sea level make it extremely unstable and totally unsuited and unsafe for development. In fact, the island is no more than a beach migrating over salt marsh along much of its length. Only the west end consists of a small land area, but it too is eroding. Overwash is extreme. Erosion is extreme. The entire island lies in the V-zone. The overall risk is extreme and will remain so because the island is on the move. Marsh mud and peat are exposed over most of the beach, along with a few stumps that give evidence of a long-lost maritime forest and the rapid shoreline retreat. Tidal creeks and marsh areas shift frequently so that even recent maps are inaccurate. Inlets frequently open and close, and at one time this sand strip was part of Botany Bay Island to the east.

Risk Zones

:::::: Low

▨ Moderate

▨ High

▨ Extreme

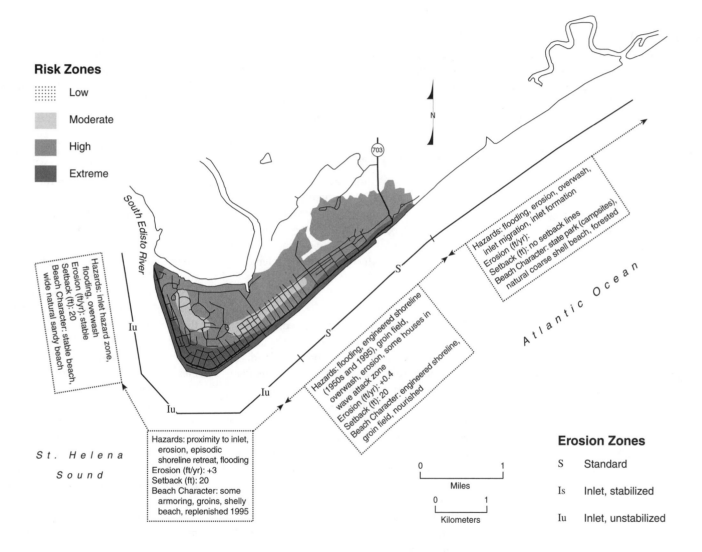

South Edisto River

(703)

N

Atlantic Ocean

Hazards: flooding, erosion, overwash, inlet migration, inlet formation
Erosion (ft/yr):
Setback (ft): no setback lines
Beach Character: state park (campsites), natural coarse shell beach, forested

S

Iu

Iu

S

Hazards: inlet hazard zone, flooding, overwash
Erosion (ft/yr): stable
Setback (ft): 20
Beach Character: stable beach wide natural sandy beach

Iu

Hazards: flooding, engineered shoreline (1950s and 1995), groin field, overwash, erosion, some houses in wave attack zone
Erosion (ft/yr): +0.4
Setback (ft): 20
Beach Character: engineered shoreline, groin field, nourished

St. Helena Sound

Hazards: proximity to inlet, erosion, episodic shoreline retreat, flooding
Erosion (ft/yr): +3
Setback (ft): 20
Beach Character: some armoring, groins, shelly beach, replenished 1995

0		1

Miles

0		1

Kilometers

Erosion Zones

S   Standard

Is   Inlet, stabilized

Iu   Inlet, unstabilized

RM 15. Edisto Island and South Edisto River

Individual Island Analyses   117

Edingsville Beach (RM 15)

What remains today of Edingsville Beach is bounded by Frampton Inlet and Scott Creek, and backed by salt marsh. Eighteenth-century maps indicate that the island once had beach ridges, but today it is a low, narrow, migrating barrier island with severe overwash and rapid erosion. The front side has eroded at rates of up to 100 feet per year, and exposed marsh sediment is common on the front side of the island.

The spectacular nature of beach retreat on this island is vividly illustrated by the fact that the small town of Edingsville Beach, once located on the central portion of the island, is now a ghost site offshore! Prior to the "Sea Islands Hurricane" of 1893 and before the days of hurricane warnings, Edingsville was a summer resort for the plantation gentry, and a good alternative to malaria. But that storm's rising waters flooded the causeways and evacuation routes first. The storm surge destroyed the town and erased the island together with 3,000 souls from the sea islands. Today nothing is left of the town or the island except a washover terrace and an occasional polished brick found on the beach.

The location of Edingsville Beach on the shore could not persist in the face of the rising sea level. Likewise, beach ridges and maritime forest were no guarantee that the island would be stable even for 100 years. Few coastal property owners know of Edingsville Beach and similar sites lost beneath the waves, but if the old town's residents could come back, they could teach us much about imprudent coastal development. Nothing stays the same on a barrier island, including the island's position. Today this island certainly is an extreme-risk zone unsuitable for development.

Colleton County (RM 15)

The secession of Edisto Beach from Charleston County reshaped the county line and gave Colleton County an ocean shoreline consisting of that single barrier island. The remainder of the county's coast is made up of Pine and Otter Islands on the northeast side of St. Helena Sound.

Edisto Beach (RM 15)

The name Edisto Beach (fig. 6.26) designates the outer, barrier island portion of the larger Edisto Island complex. This barrier island, along with Edingsville Beach and Botany Plantation, is separated from the main body of Edisto Island by tidal creeks and salt marshes. Edisto Beach changes character from its northeastern end, where it is more like its narrow, overwashed, and eroding sister islands, to its southern portion, where it widens into forested beach ridges with moderate elevation.

Edisto Beach State Park (RM 15) occupies the northern one-third of the 4-mile beachfront. Recreational use is a good choice for this stretch of island because it is not suitable for development. The proximity to Jeremy Inlet and the record of erosion, overwash, and flooding mark the area as an extreme-risk zone. A walk on the park's beach shows an abundance of coarse oyster-shell debris mixed with fossils.

Edisto Island has a beach unlike any other in South Carolina. It is composed of coarse shell hash, and in the nature of large sediment beaches it is relatively steep. Why is this beach so unusual and where does the shell hash come from? It has several sources.

It is possible to find on the beach remnants of the former town of Edingsville in the form of polished brick and pottery fragments. But there is not sufficient material from that source to cover Edisto's beach. The fossils mixed in with the shell debris indicate that the formations being eroded along Edisto date back to the early Pliocene, which began about 10 million years ago. This is the source of the great shark teeth so prized by collectors. Younger, Pleistocene fossils are present as well, including the remains of horses, camels, mastodons, marine mammals, and smaller sharks. Mixed with these ancient remains are a vast number of contemporary invertebrate shells.

By far the greatest portion of Edisto's shell hash beach was put there by humans. Following the destruction wrought by Hurricane Gracie in 1959, the Army Corps of Engineers gave Edisto a new front beach. A large area of marsh landward of the island was excavated, and the contents were pumped to the shoreline. The source came to be called the Edisto

6.26. Groin field on the southwest end of Edisto Island, at the point. Note the narrow dry beach and contrasting setbacks of houses. In 1995 a small beach replenishment was carried out on Edisto Beach, but chances are good that the beach will disappear within a few years. The cost of attempting to protect this property will continue to escalate.

ishment project not only restored the lost beach but also created highly desirable—and highly erosional, front-beach property. Where, oh where, is the lesson of Edingsville Beach?

The developed middle portion of the island is designated a moderate-risk zone. Individual site safety is highly variable over most of the island. Much of Edisto Beach has a front row of houses facing the ocean with little protection (extreme risk). Some houses have a protective dune line; many do not. These are subject to flooding and wave velocity should hurricanes strike. And strike they have. The 1940 hurricane destroyed 175 cottages on Edisto Beach and cut back the dunes 30 to 120 feet. Some reports indicate that the central shore portion of the island is stable, but groin fields are constructed where erosion is or has been a problem. The groin field itself should raise a flag of caution. Frontal houses have been damaged or destroyed since 1940, and will be again. Avoid beachfront property. Seek sites on high ground (elevations on the island go as high as 30 feet) and where there is protective forest cover and frontal dunes.

Yacht Basin. We have found no record of the percentage of sand in this immense volume of sediment. It is certain, however, that for years a plume of marsh mud was washed from the new beach and spread as a lethal blanket on the ocean floor, the foundation of the marine ecosystem. Eventually all that remained on the beach was the sandy shell hash.

Two final points may add perspective to the story of the artificial shell beach. Today there are federal (EPA) and state (OCRM) laws to prevent this kind of environmental destruction, whether it be done in ignorance or in spite of knowledge. However, protection of the environment still rests largely with the will of the community. Second, prior to Hurricane Gracie, no developed properties existed seaward of Palmetto Boulevard. The great replen-

The interior portion of the south end of the island is classed as a high-risk zone, but here again, lower-risk sites are found toward the higher-elevation inner part of the island. Avoid beachfront property, even where accretion is occurring, as it is along parts of the Edisto Beach shore. If you own beachfront property that is accreting, encourage dune formation in front of your property by planting dune grass and building sand fences.

The development of Edisto Beach dates back to the early part of this century. For decades the community had a unique flavor all its own, and growth was relatively slow. Edisto Beach is probably now at a turning point in its development history. The island's population is likely to double in the next decade, doubling the demand for water and doubling the septic waste load, taxing nature's ability to absorb without endangering human health or damaging wildlife. The new resort development may reflect a turn away from the traditional beach cottage residences.

In the early 1970s, an Army Corps of Engineers report concluded that there should be no federal participation in an erosion control project for Edisto Beach State Park because an adequate source of sand available at a reasonable price was lacking. In 1995, state and local funds were used to construct a minimum size beach replenishment project. This was done essentially to buy some time for the erosional properties alone the northern half of Edisto Beach. Precious little time was purchased.

Edisto Beach is a residential island rather than a tourist beach, and property rentals and sales here do not generate the enormous amount of money needed to hold back a retreating shoreline. Perhaps future funds would be better spent moving structures rather than trying to protect them. It is just a matter of time. If quaint Edisto Beach is the first to bite the bullet, it will not be the last.

Pine Island and Otter Island

These two islands making up the sound coast were not included in the risk maps. They lie between the South Edisto River and St. Helena Sound, and both are primarily salt marsh; the little high ground is less than 10 feet above mean sea level. Both islands are privately owned and are characterized by small maritime forests and scrub shrub communities. The islands are low lying, inaccessible by land, and vulnerable to storm surge. They should be viewed as an extreme-risk to high-risk zone unsuitable for development.

The Sound Islands (RMs 16 and 17)

St. Helena Sound and Port Royal Sound are the boundaries of a group of pristine, isolated sea islands, many still undeveloped (fig. 6.27). Risk maps were made for the developed portions of these islands.

Beaufort County (RMs 16, 17, 18, and 19)

This coastal segment is characterized by large estuaries separating low-lying coastal plain headlands fringed by barrier islands. The islands vary from large complexes of old Pleistocene (Ice Age) islands onto which younger marshland and modern barrier systems have attached (e.g., Hilton Head Island), to small, transgressive islands (e.g., Little Capers Island) characterized by rapid erosion and extensive overwash. The large tidal range and extensive marshes associated with the large sounds cause strong tidal currents and complex patterns of erosion and deposition. The smaller islands between inlets are generally unstable and must be classified as extreme-risk to high-risk areas. In some cases erosion rates are extreme and overwash potential is severe. In addition, the flood hazard is significant, and there may be problems with storm evacuation. Consistently, the moderate-risk sites are on the higher, forested dune ridges away from the beachfront.

People who live on or who are seeking coastal property on the sounds (e.g., on St. Helena Island) should evaluate their sites and structures just as if they were locating on oceanfront property. The ocean islands from northeast to southwest are Harbor Island, Hunting Island, Fripp Island, Pritchards Island, Little Capers Island, St. Phillips Island, Bay Point Island, Hilton Head Island, and Daufuskie Island.

This region was the scene of the August

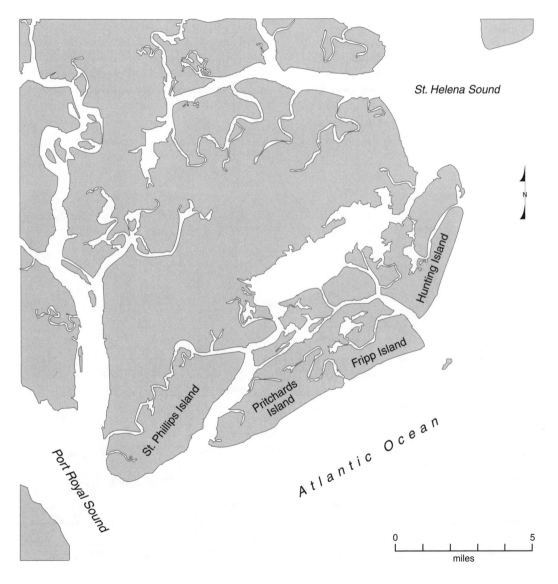

St. Helena Sound

Hunting Island

Fripp Island

Pritchards Island

St. Phillips Island

Port Royal Sound

Atlantic Ocean

N

0   5
miles

1893 "Sea Islands Hurricane" that destroyed Edingsville Beach. Jeff Rosenfeld, managing editor of *Weatherwise* magazine, has called it the "forgotten" hurricane (appendix D, ref. 18). Why forgotten? It is forgotten in the sense that few current coastal residents know of the thousands of lives lost, the tens of thousands displaced, and the ruined economy it left behind. It is also forgotten in terms of lessons unlearned as the sea islands are now being occupied by development of increasing density that puts more and more people and property at risk. A repeat of the 1893 storm will again raise water levels to nearly 20 feet above mean sea level, will again send waves crashing over the islands, and will again blow winds at 120 miles per hour. The price tag this time will be in the multibillions, and the suffering and loss will be not unlike that experienced on August 28, 1893.

Harbor Island (RM *16)*
Harbor Island lies behind Hunting Island across Johnson Creek (fig. 6.28); however, the northeast end of the island faces the open water of St. Helena Sound. Harbor is a transgressive island formed of sand transported up St. Helena Sound (fig. 6.27). Interestingly, Harbor Island invariably benefits from the renourishment of Hunting Island.

6.27. Index map of the Sound Islands, the small islands located between St. Helena and Port Royal Sounds.

Finger canals have been dredged at the southern end of the shoreline. Although small patches of maritime shrub forest exist behind a well-developed frontal dune, the area is one of extreme to high risk. New condominiums at the marsh edge are within the flood zone, and the erosion rate has been measured at greater than 15 feet per year (appendix D, ref. 49). The orientation of the beachface makes it susceptible to erosion from northeasters. This development should be viewed with great caution.

Hunting Island (RM 16)
Hunting Island has a replenished beach, but it is a state park, close to being in its natural state, and it will not see extensive development (fig. 6.27). Beach nourishment has attempted to maintain the island's recreational beach. The first beach nourishment took place in the late 1960s when 500,000 cubic yards of sand were placed on the beach. By the end of 1970 this sand had all but disappeared, and a second beach nourishment project dumped another 750,000 cubic yards of sand in 1971. Within six months, 97 percent of the second nourishment had been lost to erosion (appendix D, ref. 80).

In part, the rapid loss may be accounted for because the borrowed sand included fine dune sands from the westerly side of the island, and this eroded faster than the coarser natural beach sand. In addition, the island is only 4.3 miles long and bounded by inlets. The result-

ing dynamics produces strong currents that remove sand, particularly during storms. The erosion rate is approximately 300,000 cubic yards per year. From 1857 to 1979 the north half of the island retreated between 985 and 2,460 feet (appendix D, ref. 53). An important lesson was learned: not only was sand lost from the beach, but also from the island, the source of the nourishment. Such cannibalism cannot be maintained.

The third beach nourishment came in 1975 when another 613,000 cubic yards of sand was added, this time borrowed from the shoals at the mouth of Fripp Inlet. The total cost of these first three nourishments was just over

6.28. Odd beach accretion pattern at Harbor Island, constructed of sand lost from the beach replenishment project on Hunting Island.

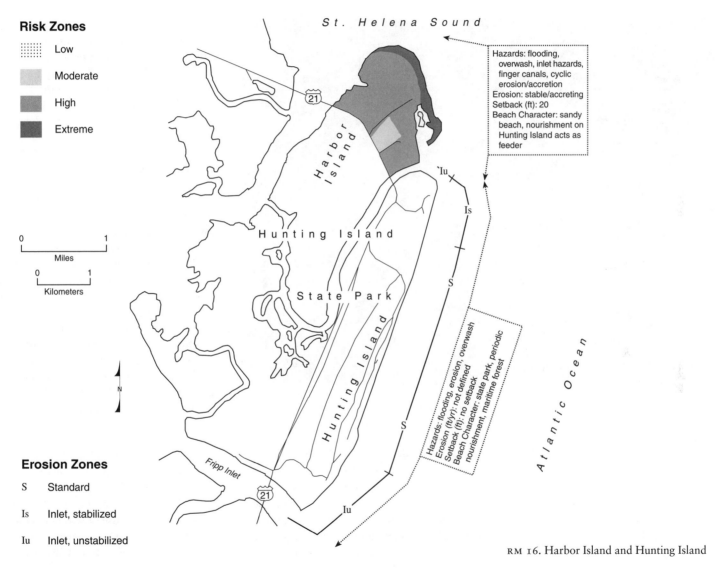

Risk Zones

▦ Low

▨ Moderate

▨ High

■ Extreme

St. Helena Sound

Hazards: flooding, overwash, inlet hazards, finger canals, cyclic erosion/accretion
Erosion: stable/accreting
Setback (ft): 20
Beach Character: sandy beach, nourishment on Hunting Island acts as feeder

H a r b o r   I s l a n d

H u n t i n g   I s l a n d

S t a t e   P a r k

H u n t i n g   I s l a n d

Hazards: flooding, erosion, overwash
Erosion (ft/yr): not defined
Setback (ft): no setback
Beach Character: state park, periodic nourishment, maritime forest

Atlantic Ocean

Fripp Inlet

0 _____ 1
Miles

0 _____ 1
Kilometers

N

Erosion Zones

S   Standard

Is   Inlet, stabilized

Iu   Inlet, unstabilized

RM 16. Harbor Island and Hunting Island

merly uneroded areas behind the beach.

Today, between renourishments, you can see fallen trees and wave-cut scarps in the dunes at the back of the beach (fig. 6.29). Severe long-term erosion continues (appendix D, ref. 113), but no property is threatened other than the park road and a few cabins. Yet the response to the appearance of stumps on the beach as this young island (perhaps no more than 5,000 years old) attempts to migrate is to rebury them and maintain an artificial beach at high cost. We have not learned the lesson of our forefathers who constructed the present lighthouse in 1875 so that it could be disassembled and moved back when threatened by erosion. Such a move was necessary in 1889. The capacity to move with the moving shoreline is less expensive than defending the structure against the forces of nature!

Hunting Island remains an excellent public recreation area where you can enjoy not only the beach but the maritime forest and marsh as well. The climb to the top of the lighthouse is an effort rewarded by a magnificent view of the island, associated inlets, and tidal deltas.

6.29. Lighthouse on the northern end of Hunting Island. The erosive shoreline in this view is in between nourishment projects. There have been at least four major replenishment projects here; all disappeared quickly.

$2.1 million, of which the state paid $660,000. By 1976 stumps were reappearing on the beach, and by 1977 the next nourishment project was being planned (appendix D, ref. 118); more recently, 1.4 million cubic yards of sand was placed in 1980. The total cost of these four projects was well over $4 million, and renourishment probably will be required every three years (appendix D, ref. 79).

The most recent renourishment, performed in 1990, had a design life of three years. The design was accurate. As of this writing the renourished sand is gone and the 150 feet of shoreline retreat since 1990 has penetrated for-

Fripp Island (RM 17)
Three-mile-long Fripp Island lies southwest of Fripp Inlet and is accessible by bridge from Hunting Island (figs. 6.27, 6.30). Its short length makes much of the island unsuitable for development because of strong inlet influence (appendix D, ref. 119). However, Fripp Island is well over a mile wide at its widest point and

has forested dune ridges with elevations to 25 feet. Unfortunately, development has not been restricted to these lower-risk sites.

The sheltering effect of the ebb-tidal delta off Fripp Inlet has resulted in beach accretion on the northeastern one-third of the island. To the southwest the measured erosion rates increase (appendix D, ref. 119). Even in the zone of accretion, rapid short-term erosion can be severe (e.g., in 1955–1960), and the short existence of the accreted land adjacent to the inlet has not allowed further stabilization through the buildup of dunes or growth of maritime forest. The developers did not take advantage of the natural protection that this buffer zone would have provided and have encroached onto the low-lying area adjacent to both ocean and inlet. A terminal groin and attached seawall and rock groins were constructed to protect the property. Similarly, groins have been constructed on the south end of the island. Shoreline engineering structures (fig. 6.31) were chosen as the course of action in the initial early development, so even though the Fripp development is modern, it does not follow the philosophy of designing to live with nature. People seeking to minimize risk on Fripp Island should consider locations on the forested dune ridges of higher elevation

6.30. Aerial view of Fripp Island. Note the narrow to nonexistent beach in front of the seawall. A stiff environmental price, loss of the beach, has been paid to save beachfront buildings.

in the interior. The exception is where such ridges terminate at the inlet channel and are being eroded. Avoid beachfront property. Keep in mind that the escape route from this island is onto another island and then across extensive lowlands. In the case of a hurricane warning, evacuate early.

Pritchards Island (RM 17)
Pritchards Island is a small (2.5 miles long), undeveloped island that lies between Skull Inlet and Pritchards Inlet (fig. 6.27). It is the site of the coastal research facility of the University of South Carolina at Beaufort. Erosion along much of the oceanfront is moderate to severe, and the beach penetrates into the edge of the forest. Trees litter the beach, and the absence

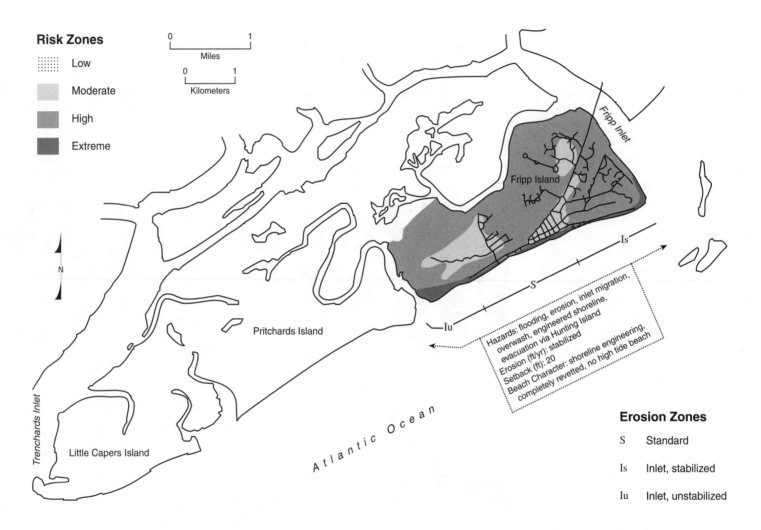

Risk Zones

:::::: Low

▒ Moderate

▓ High

█ Extreme

0 ——— 1
Miles

0 ——— 1
Kilometers

N

Fripp Inlet

Fripp Island

Is

S

Iu

Hazards: flooding, erosion, inlet migration, overwash, engineered shoreline, evacuation via Hunting Island
Erosion (ft/yr): 20
Setback (ft): stabilized
Beach Character: shoreline engineering, completely revetted, no high tide beach

Pritchards Island

Trenchards Inlet

Little Capers Island

Atlantic Ocean

Erosion Zones

S   Standard

Is   Inlet, stabilized

Iu   Inlet, unstabilized

RM 17. Fripp Island

of a frontal dune line allows overwash. Forested dune ridges on the interior of the island with elevations of 10 feet provide a limited area for potential development (high hazard), but much of the area must be viewed as an extreme-hazard zone. It is not likely that a bridge will be built to the island.

Little Capers Island

Little Capers Island, only 2.5 miles long, has a severe erosion problem (greater than 95 feet per year in some places), and extensive exposures of old marsh mud and peat are common on the beach. The proximity of Pritchards Inlet and Trenchards Inlet to the west account for the instability of this low marsh island (fig.

6.27). Overwash is widespread and severe (appendix D, ref. 119). The few isolated dune ridges are less than 10 feet in elevation and are covered by remnants of maritime forest. The ridges run perpendicular to the trend of the beach, and their ends are eroding, as indicated by wave-cut scarps and fallen vegetation. Of the few fishing cottages on the island, at least three are in the oceanfront extreme-hazard zone (fig. 6.32). There is no bridge to the island, and development is regarded as unwise.

St. Phillips Island

This large, privately owned island lies in back of the barrier formed by Little Capers Island and Pritchards Island, forming the inland

6.31. Left: Fripp Island revetment. No recreational dry beach exists in front of this 20-foot-high rock wall. The vegetation debris about halfway up the wall is from a recent normal high tide.

6.32. Right: Fishing shack on the beach at Little Capers Island. The beach is retreating into the forest. Contrast this natural retreat, where beach continues to exist, with retreat on Fripp Island (fig. 6.30), where a wall blocks retreat and the beach was destroyed.

shore of Trenchards Inlet, a shifting tidal channel (fig. 6.27). The southwestern end of the island is behind Bay Point Island; however, a short stretch (less than 1.5 miles) fronts the open ocean at the mouth of Trenchards Inlet. This beachfront is highly unstable and severely erosional, with exposed stumps and marsh sediment as well as fallen trees. In addition, its flood potential, its position within the V-zone, and the lack of escape routes make it an extreme-risk zone.

Overall, the island is not suited for development. Inland the island is composed of narrow, forested dune ridges, but even these ridges are low (less than 10 feet), and the entire island would be inundated by a flood of less than a 100-year level. The extensive marsh between ridges, intricate tidal creeks, and distance to safe havens inland are barriers to safe escape in the event of evacuation.

There are no current plans to develop this island, which is accessible only by boat. Almost certainly a bridge would lead to development that would exceed the island's carrying capacity. In addition, access to St. Phillips Island would increase the pressure to extend development to adjacent Capers Island, Pritchards Island, and Bay Point Island, most of which are in extreme-risk zones that are unsuitable for even limited development. St. Phillips Island was purchased in 1972 for $906,000. By 1978 the asking price was $2.5 million (appendix D, ref. 52). The opportunity for inflated profits is certainly a driving force in developing islands such as these.

Bay Point Island

This small island, dominated by marshland, lies between Trenchards Inlet and Port Royal Sound (fig. 6.27). With only 1.3 miles of beachfront, distant access even by boat, and a highly unstable shoreline, it is not likely to be developed. The island is privately owned, and there are some sites on forested high ground away from the beach that might be viewed as suitable for development. Overall, however, the island is an extreme-risk zone.

At the beginning of this century Bay Point Island was not separated from St. Phillips Island, because Morse Creek did not have an outlet to the Atlantic. By 1919, however, the St. Phillips shoreline had eroded to the point that Morse Creek opened into the ocean, forming Bay Point Island. The fate of Confederate Fort Beauregard, once located on the island but now underwater, testifies to the long-term erosion problem.

Hilton Head Island (RM 18)

Hilton Head Island is one of the cradles of modern coastal development. The planned development of this large sea island began in 1956 with Charles Frazier's visionary Sea Pines Plantation. It has since been transformed from a natural and rural area into a complex of coastal resorts and suburbs. The population grew from 1,000 in 1960 to about 10,500 in 1980 (appendix D, ref. 79), and the 1995 population exceeded 27,000 (appendix D, ref. 7). There has been a corresponding rise in property values. Hilton Head is a model, both good and bad, that has been imitated by numerous coastal developments along the eastern and Gulf seaboards.

At 11.5 miles long and 6.8 miles wide, Hilton Head is the largest barrier island on the South Carolina coast (fig. 6.33). The part of the island southwest of Folly Creek is a Pleistocene (Ice Age) beach-ridge island. The ridges are well forested and often exceed 20 feet in elevation (these are good low-risk sites). Northeast of Folly Creek is a younger transgressive island growing in much the same way as Harbor Island, by sediment moving landward, up the estuaries.

As shown on Risk Map 18, the north end of the island, adjacent to Port Royal Sound, is an extreme-risk to high-risk zone. Erosion in this area has been significant, with measured losses of up to 17 feet per year (appendix D, ref. 118) and a total retreat of between 650 and 975 feet between 1898 and 1977 (appendix D, ref. 119). However, the houses here are set well back, and no habitable structures are threatened (fig. 6.34).

The ocean side of the northern half of the island is also experiencing long-term erosion, with rates of loss up to 6 feet per year. The area between Forest Beach and about 4 miles north of Folly Creek including Burkes Beach (fig. 6.35) has retreated by as much as 1,600 feet since 1898; the maximum erosion occurred along a 5-mile section south of Folly Creek where the small tidal inlet is also migrating (appendix D, ref. 79). Although the area in front of Forest Beach was accreting in the early 1900s, the recent trend has been erosive retreat, especially since the early 1970s. Studies

RM 18. Hilton Head Island and Calibogue Sound

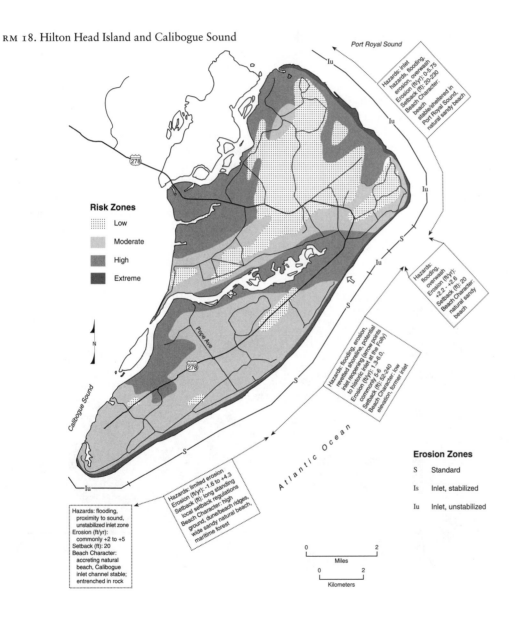

Port Royal Sound

Hazards: inlet hazards, flooding, erosion, overwash
Erosion (ft/yr): 0-5-75
Setback (ft): 20-230
Beach Character: stable/sheltered in Port Royal Sound, natural sandy beach

Risk Zones

Low
Moderate
High
Extreme

Hazards: flooding, overwash
Erosion (ft/yr): +2.2 - +2.6
Setback (ft): 20
Beach Character: natural sandy beach

Pope Ave

Hazards: flooding, erosion, reverted shoreline, potential inlet reopening (arrow points to historic inlet at the Folly)
Erosion (ft/yr): 1.3-6.0, commonly 5-6
Setback (ft): 52-240
Beach Character: low elevation, former inlet

Calibogue Sound

N

Atlantic Ocean

Erosion Zones

S — Standard
Is — Inlet, stabilized
Iu — Inlet, unstabilized

Hazards: limited erosion
Erosion (ft/yr): -1.6 to +4.3
Setback (ft): long standing local setback regulations
Beach Character: high ground, dune/beach ridges, wide sandy natural beach, maritime forest

Hazards: flooding, proximity to sound, unstabilized inlet zone
Erosion (ft/yr): commonly +2 to +5
Setback (ft): 20
Beach Character: accreting natural beach, Calibogue inlet channel stable; entrenched in rock

0   2
Miles

0   2
Kilometers

of wave patterns show wave energy being focused on the Forest Beach section (appendix D, ref. 79).

The residents' response to the erosion was predictable. Houses constructed too close to the ocean were threatened. A hodgepodge of stone seawalls, nylon sand bags, and beach scraping from foreshore to backshore was used to combat the erosion, with the actual effect being possible acceleration of erosion and the destruction of the aesthetics of the beach. The small size of the rocks in the revetments increases the likelihood that they will be missiles of destruction during a hurricane; it is as if the beach has been lined with small cannonballs.

A beach renourishment project was performed along the Atlantic shore of Hilton Head Island in 1990 with a 6–8-year intended project life. The sand has held up well, owing partly to the fact that no storms have hit South Carolina since then. Nevertheless, rocks are once again exposed along North Forest Beach.

The frontal part of the northern half of the island faces other hazards such as flooding (low elevations with poor natural protection), overwash, and formation of new swashes. Most of the area is classed as extreme to high risk for beachfront property.

The southern half of the island benefits from the losses farther north. Eroded sand moves south in the longshore drift and accretes to form a wider beach-dune system (fig. 6.36). As the erosion and overwash potentials are reduced, the risk goes down. This area is in the flood zone, however, and the best sites are on well-vegetated high ground away from the beach. Note that the immediate south end of the island is an extreme-risk zone because of its location next to Calibogue Sound and the resulting high erosion rate and flood potential, but the risk grades through high to moderate as you go inland.

The new bridge to Hilton Head Island is symbolic of another problem typical of barrier-island development: the need to constantly enlarge support systems until the carrying capacity of the island is exceeded. One emerging crisis on Hilton Head is an inadequate supply of high-quality water. Saltwater intrusion into the limestone aquifer is already a problem, as is the disposal of partially treated wastewater, much of which is used to irrigate golf courses. Also, as the population grows, the evacuation problem increases (even with big bridges). In each instance the quality of the natural environment suffers, until finally socioeconomic problems become more serious than natural hazards.

Daufuskie Island (RM 19)
As the southernmost of the South Carolina sea islands, Daufuskie has been one of the last remaining plums ripe for development. Bounded by the Cooper and New Rivers and Mungen Creek, and separated from the Savannah River by low, marshy islands and shoals, Daufuskie Island remained isolated through much of its history. Calibogue Sound, the northeast

6.33. Below: Left: Vertical aerial photograph of Hilton Head Island, actually two geologically distinct islands separated by Folly Creek. Most of the forested areas have since been developed. *Source:* NOAA Infra-red High Altitude Photography (INHAP), 1980.

6.34. Right: Old steam gun emplacement on Hilton Head Island facing Port Royal Sound. Though this area is prone to long-term erosion, no habitable structures are threatened.

boundary of Daufuskie, separates it from Hilton Head Island, which until recently was a world away.

Daufuskie Island stands as an example of a barrier island that has supported a reasonable population for a long period because the island's natural system has not been overwhelmed (fig. 6.37). The island is more than 2.5 miles wide, with considerable high ground and good maritime forest cover. In the mid-1970s only about 200 persons lived on the island, but earlier in the century the number was higher, more than 700 (appendix D, ref. 52). The longtime residents were the descendants of slaves, but the island's future seems to be following that of Hilton Head as the black population sells the land to developers.

Historically, storm-surge flood levels here have approached 20 feet. The island front (there is no actual beach) would be in the wave and overwash zone at such a depth. Although no detailed studies exist, the currents around the island appear to flush away eroded sediment without replacing it by onshore drift. Erosion rates have been highest near the ends of the island. The quality of the offshore waters at the south end of the island is reported to be poor because of pollution, probably from the Savannah River.

The margins of this island are quite erosional, and the setbacks are measured in hundreds of feet. To the credit of its developers, Daufuskie is being developed accordingly. The

best sites will be those with elevations above 20 feet in the interior of the island, where the maritime forest has been maintained. Access to Daufuskie Island is by boat only. A prudent response to storm warnings is called for: evacuate at the earliest warning.

6.35. The aesthetics of Burkes Beach, Hilton Head Island, are destroyed by this rubble revetment. This beach could be refurbished and restored to a more natural state by removing the revetment.

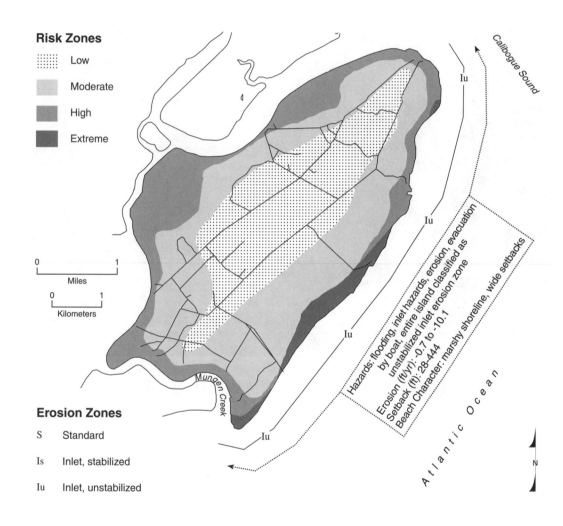

Risk Zones

:::::: Low

Moderate

High

Extreme

0   1
Miles

0   1
Kilometers

Erosion Zones

S   Standard

Is   Inlet, stabilized

Iu   Inlet, unstabilized

Callibogue Sound

Iu

Iu

Iu

Iu

Iu

Mungen Creek

Hazards: flooding, inlet hazards, erosion, evacuation by boat, entire island classified as unstabilized inlet erosion zone
Erosion (ft/yr): -0.7 to -10.1
Setback (ft): 28-444
Beach Character: marshy shoreline, wide setbacks

Atlantic Ocean

N

RM 19. Daufuskie Island

Jasper County

The southernmost coastal county of South Carolina is bounded by 2.8 miles of undeveloped shoreline (fig. 1.1). The low, marshy islands are extreme-hazard zones and are not likely to be developed. Jones Island lies along the state line behind Turtle Island and adjacent to the Savannah River. A small extension of the marsh island fronts the ocean. A portion of the island adjacent to the river was used for dredge-spoil fill in the past, and the built-up area has been suggested as a possible site for industrial development, depending on river port access.

Turtle Island is owned by the state and managed as a wildlife refuge. It is approximately 2.5 miles long and consists of a narrow beach ridge backed by salt marsh. Elevations are less than 10 feet above sea level, and the island is entirely within the flood zone (V- and A-zones). The north end of the island is prone to rapid erosion, whereas the south end has grown slightly seaward.

Past Reflections: Future Expectations

Locating at the shore calls for greater prudence than is necessary in selecting a site inland. First, the environment at the shore is less uniform. Topography, sediment, and vegetation change abruptly from beach to dune to overwash terrace to forest to marsh. These systems are not stable for very long periods, and they can change or shift in response to the rise

in sea level and shoreline migration. Look to the future when selecting a site. The future comes sooner on barrier islands. And in the words of George F. Will, "The future has a way of arriving unannounced."

Second, coastal hazards such as hurricane winds, storm-surge flooding, wave attack, overwash, and persistent erosion are unlike anything you've experienced in the relative quiet and stability of the coastal plain or piedmont. Choose your site with such forces in mind, and reinforce your construction to improve its strength to survive these forces. Expect maintenance to be more frequent and somewhat more expensive.

Third, coastal communities have their own set of dynamics, both because they are resorts and because they must respond collectively to coastal hazards. Responses to erosion, flooding, overwash, and other hazards, as well as the kind and extent of development allowed, set a course in coastal communities that is more difficult to alter than a course in an inland town. Expect increasing regulation. This pattern is suggested in the next chapter.

6.36. Left: The Sea Pines Plantation area at the southwestern end of Hilton Head Island. Though narrow, this beach is utilized for recreation and is currently building seaward.

6.37. Right: Daufuskie Island has no sandy beaches, but the beautiful interior forest is attractive for plantation-style development. Photo courtesy of Fritz Aichele.

In addition to the storm hazards we have already discussed, South Carolina's coastal zone is subject to potential dangers of the generic sort, such as earthquakes, fires, and floods. The mitigation of such hazards begins with a basic understanding of the processes involved in them. Communities and property owners should anticipate calamities and take actions to reduce their effects.

On October 17, 1989, less than one month after Hurricane Hugo and scarcely an hour before the first pitch of game 3 of the 1989 World Series, another natural disaster struck the United States: the Loma Prieta, California, earthquake. Twin disasters on opposite coasts, both devastating in terms of personal loss and cost. Along with Hurricane Andrew in 1992 and the great Midwest river floods of 1993, these events demonstrated that multiple natural disasters can generate economic and social catastrophes when they occur simultaneously (12 insurance companies failed as a result of Andrew). A longer-term view may show the 1990s as a turning point in the way our society plans for natural disasters.

The problems generated by multiple natural hazards are being recognized globally as well, and the decade of the 1990s was designated the International Decade of Natural Hazards Reduction. A global effort is under way to increase awareness, planning, preparation, response, and mitigation of all natural hazards. Along with the international effort, several nations created their own programs with blue ribbon panels and committees of experts volunteering their time, efforts, and knowledge to address the problems generated by natural hazards. The U.S. Decade of Natural Hazard Reduction, sponsored by the Federal Emergency Management Agency (FEMA) and several other federal agencies and coordinated through the Natural Hazards Reduction and Application Information Center in Boulder, Colorado (see appendix B for address), is part of the international effort. Appendix D lists several publications already available from these programs.

The multihazard approach is appropriate because the planning, preparation, and response designed to mitigate one hazard overlap considerably with mitigation procedures for other hazards. For instance, a house built to withstand earthquakes is also hurricane resistant. Likewise, evacuation and sheltering plans for hurricanes are essentially the same as those used in the case of large riverine floods. Individuals who are prepared for a hurricane are better prepared for an earthquake as well. Appendix A presents a set of hazard safety checklists. Use it when the situation arises, but read it beforehand, too, so that you will be prepared for emergency situations.

Earthquake Risk

In the great stretch of barrier islands from Long Island, New York, down around Florida to Texas, South Carolina has a unique distinguishing feature: its coastal zone is also at high risk from earthquakes. In the summer of 1886, Charleston had nearly recovered from the "great cyclone" (hurricane) of August 1885. The storm had destroyed the battery wall and seriously damaged the first row of houses behind it. Day-to-day activities helped to blunt the painful memories of the storm, just as they had helped over the 20 years since "the recent unpleasantness" that had set the North against the South. Each year brought more news, people, commerce, and ships to the harbor of the Holy City. The square-riggers were being replaced by the new steam-driven paddlewheelers. Commerce spurred development and economic growth. An aura of excitement surrounded the construction of the jetties at the harbor entrance. Charlestonians speculated that this improvement to navigation would mean a great deal to Charleston and to reconstruction in the South. Everything seemed fine on the evening of Tuesday, August 31. It was a warm evening, as it always is in August.

The earthquake struck just before 10:00 P.M. The loud, eerie moan that preceded it brought many people out of their houses, some already in their nightclothes, and then the world as they knew it turned upside down. The ground heaved, and masonry buildings collapsed (fig. 7.1). Fires ignited and raged out of control. In the surrounding countryside the people witnessed a bizarre phenomenon: the landscape was dotted with "volcanoes" spewing sand and water into the air (fig. 7.2). Even grizzled

7.1. It was a bad day for business at Bird's Oil and Paint Store at East Bay and Cumberland Streets the day after the 1886 Charleston earthquake. Many more of Charleston's older buildings could collapse in a future earthquake, and modern buildings are not immune from a similar fate. Photograph courtesy of the South Carolina Historical Society.

war veterans had seen nothing to prepare them for this.

Of all the earthquakes recorded in the United States, only a few have exceeded the 1886 Charleston event in magnitude and dam-age. The dead numbered 110; 27 died immediately as a result of collapsing buildings and falling debris, and the death toll rose as people died from injuries and ensuing disease. Damage to buildings was estimated at $6 million, one-fourth the total value of the city's buildings. More than 2,000 buildings were damaged or destroyed, leaving thousands homeless. Twelve thousand chimneys were destroyed. (Many of the surviving buildings were retrofitted with rod-bolts. The decorative ends of these stabilizers are still visible on many Charleston buildings, as are a few cracks attributed to the earthquake.) The newly built Morris Island lighthouse (Charleston Light) received surficial cracks that are still visible today.

The magnitude of the Charleston earth-quake is estimated to have been 6.5 to 7.0 (a great earthquake) on the modern Richter scale (IX–X on the Mercalli scale, indicative of almost complete destruction and collapse of buildings). Unlike West Coast quakes, major earthquakes generated in the Charleston Seismogenic Zone travel long distances and cause damage hundreds of miles from the epicenter (fig. 7.3). The quake shook the earth in New York, rattled buildings in St. Louis, produced strong ground shaking in Chicago (800 miles away), and damaged a church in Indianapolis some 600 miles away, on the other side of a mountain range!

The Charleston earthquake prompted the first organized field mission of the American Red Cross. Fresh water, already in short supply before the earthquake, became a prized commodity. Tent cities were established on every available green space. Local diarist Ellen Herd recorded that "in some camps the people were in a pitiful state, in others there was constant card playing and drinking, men and women drinking to excess who had never done so before."

Beyond the obvious toll on human life and property, natural disasters have a psychological effect that is hard to appreciate. Most of those affected by the 1886 quake had never even heard of an earthquake, and certainly had not experienced such violent destruction. There was only one generally accepted explanation: Charleston had been visited by Lucifer himself.

Today's population in the same area may feel "experienced" after Hugo, but they are unprepared for a great earthquake, both physically and psychologically. Developments now stand in areas that may liquefy in the next great earthquake.

The Charleston Seismic Zone

The foremost obstacle to our understanding of faulting and earthquakes on the east coast of North America is the fact that active faults tend to be buried beneath the thick sediments of the coastal plain. Primarily, these faults are known from their distortion of formations near the surface and from seismic surveys. There is no place where the faults can actually be seen. Contrast this to the well-known San Andreas Fault of California, which can be traced at the surface in offset creeks, tree lines, roads, and even fences.

At present, three major faults are known to lie beneath the Charleston area (fig. 7.4; appendix D, ref. 105). The activity on the large Woodstock Fault is not well understood, nor is its influence on the region's other two faults. The Ashley River Fault (appendix D, ref. 104) is nearly vertical and roughly follows the trend of the Ashley River. This fracture extends into

crystalline rocks 2.5 to 5 miles below the surface. The effects of its movement are reflected in an upward bulge in the land surface in the area of Kiawah Island (appendix D, ref. 111). These distortions are surveyed periodically by the South Carolina Geodetic Survey in an effort to detect elevation changes that would indicate fault movement. The Charleston Fault (appendix D, ref. 105) nears the surface in the area of Sullivans Island. This broad, low-angle

fault dips toward the southwest to an unknown depth.

The three faults control the configuration of Charleston Harbor; the earth between the Ashley River Fault and the Charleston Fault gradually subsides or drops, explaining why the local rivers (Ashley, Cooper, Wando, and sometimes the Stono) flow into the harbor rather than southeast toward the sea.

The Ashley River Fault is active today, although only a few seismologists are aware of the microearthquakes detected at seismograph stations. For example, 24 earthquakes of magnitudes less than 2.5 were recorded on the seismic network at Charleston Southern University during 1994. The *Post-Courier* sometimes reports the larger earthquakes detectable to humans. These "little" quakes indicate that stress is being relieved along the fault. Scientists generally believe that larger earthquakes may occur when these microearthquakes cease and the strain begins to grow.

7.2. Liquefaction pit resulting from the 1886 Charleston earthquake. Ground shaking caused water-saturated soils and sediments to liquify, generating quicksand pits and "sand volcanoes." We have no direct knowledge of how barrier island sediments behaved in the 1886 event, but such liquefaction will destroy buildings in the next great earthquake. Photograph courtesy of the South Carolina Historical Society.

7.3. This train was literally washed off its tracks by the flood that resulted when a dam collapsed on Langley Pond near Aiken, South Carolina, in 1886. The dam was destroyed by the great Charleston earthquake, centered 120 miles away. Photograph courtesy of the South Carolina Historical Society.

Damage Caused by Earthquakes

Earthquakes generate five kinds of damage: tsunamis, landslides, fire, liquefaction, and shaking. Since South Carolina's coastline lies fully 50 miles from the continental shelf break, we are protected from tsunamis, or tidal waves, which are maximized by narrow-shelf settings where deep water is near the shore (more typical of the Pacific coast). We are likewise safe from landslides, given the lack of steep slopes and high relief in the state. Fire is one of the primary damages associated with earthquakes here, along with liquefaction and the ground shaking that ruptures fuel tanks and pipes and gas lines—and the water lines needed to combat the fires that result. Fire is discussed in a separate section later in this chapter.

Liquefaction

The pressure of seismic waves on groundwater produces some of the most curious and potentially damaging effects of earthquakes. This powerful force from below raises the water table upward through the coastal sands and muds, which are lifted along with it. The indi-

Earthquakes and Coastal Erosion Patterns

The three faults may contribute locally to coastal erosion. Erosion on Folly Island has long been attributed to sediment starvation caused by the harbor jetties, and there can be no doubt that this is a major factor. (The U.S. Army Corps of Engineers concedes that 65 percent of Folly's troubles are attributable to the harbor project.) But another reason for the erosion has to do with sea level. The global rise in sea level, for example, is causing some erosion on Folly. But sea level also rises in a rela-

tive sense when the land surface sinks, a process sometimes caused by seismic activity. Thus, the downward motion of the harbor fault block may also be responsible for some erosion. Folly Island is on the down-dropped side of the Ashley River Fault and may be in the process of seeking the same elevation as Charleston Harbor. This seismic motion also accounts for the uplift of Kiawah Island (appendix D, ref. 111), which was highly erosional in the 1880s but has been generally accretional in the present century, since the great earthquake.

vidual grains of sand are coated with a film of water and are not held together by grain-to-grain contact, friction, or gravity. The result is that the sand behaves like a liquid rather than a solid—hence the name "liquefaction" (fig. 7.2). Liquefaction can also be caused simply by shaking the already water-saturated sediment, and, in extreme cases, by intense shaking of dry sediment. Whatever the setting, the result is the same: the ground loses its ability to bear weight. Classic quicksand, a type of liquefaction not associated with earthquakes, can be created when water moves upward through the sediment for whatever reason; but the result is the same: sediment that lacks weight-bearing strength.

Liquefaction can have various effects depending on the near-surface ground and groundwater conditions. In the Mexico City earthquake of 1985, the ground in some places liquefied deeply enough and long enough for some buildings to tilt and partially sink. Liquefaction caused even greater structural damage in the 1964 earthquake in Nagata, Japan. This process isn't likely to cause an entire building to disappear; however, serious structural damage is the likely result of such settling and in effect is the same as total destruction. Cleanup alone is labor-intensive because the quicksand soon returns to its natural state.

The three New Madrid, Missouri, earthquakes (1811–1812) were some of the worst ever felt in this country. The Mississippi River actually flowed backward for a brief time, with

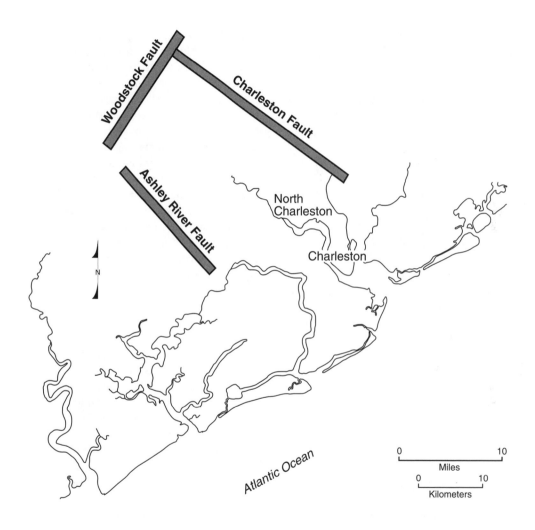

7.4. Geologic faults in the Charleston area along which future earthquakes may be expected.

one result being the creation of Reelfoot Lake. In the first of these quakes, liquefaction produced an effect different from those described above. In one small town the level of the groundwater was raised 3 to 4 feet higher than the land surface. The hapless residents found themselves wading in chest-deep water in the middle of the late-December night.

Liquefaction associated with the Charleston earthquake produced an even stranger, if perhaps less threatening, effect. The countryside was dotted with eruptions called "sand geysers" or "sand volcanoes" (fig. 7.2). These occur when seismic waves encounter lenses of saturated sand that are sandwiched between layers of impermeable sediment. The sand becomes liquefied and is ejected upward. Reports indicate that some of the sand volcanoes were 20 feet in height, though most were probably half that size. A few of the larger ones supposedly continued to erupt for most of a day. Stained sands of various colors were collected from different sites and sold or kept as souvenirs. Unfortunately for local farmers, the sand from these volcanoes was lacking in organic matter, and the new layer of sediment rendered the ground worthless for growing crops for many years. No known injuries or deaths resulted from these strange eruptions. On the other hand, a recent study of the Port Royal, Jamaica, earthquake of 1690 (appendix D, ref. 106) showed that liquefaction poses a real hazard to humans and animals caught in open, sandy areas.

The barrier islands were scarcely developed in 1886, and little was recorded of the earthquake's effects there. It is known that liquefaction of marsh muds and water-saturated island sands probably occurred. Today these unconsolidated sediments are the foundation sites of dense community development— houses, condominiums, and businesses.

Scientists are a very long way from predicting where liquefaction will occur or what form it will take. Following the 1976 Tangshan earthquake, a delegation of Chinese seismic engineers stated at a conference at Stanford University that "the problem of soil liquefaction in an earthquake is unresolved. No measures have been found to prevent damage from it" (appendix D, ref. 112).

Structures built in the coastal zone do have one advantage when it comes to liquefaction. Houses built on pilings, a requirement for structures in coastal flood zones, are thought to be as safe from this particular hazard as it is possible to be. This is because the piles normally extend below the water table, and a house supported on them provides minimum resistance to the upward force of the liquefied material. With respect to liquefaction, the primary danger to humans comes when they are caught in open, sandy areas. For buildings and other structures, the primary danger is ground motion and localized liquefaction.

Seismic Energy and Ground Motion:
The Primary Hazard

Assuming that your house does not burn as a result of an earthquake-generated fire, or that it is not dangerously close to a liquefaction site, the real danger from an earthquake is the "quake" itself. Oscillating ground motion induces an intense and complex series of strains on a structure in a very short time span. Even a minor 3.9-magnitude quake in April 1995 did cosmetic structural damage to schools in the Dorchester, South Carolina, district (e.g., cracks in walls, warping of floors, and damage to stairwells).

The science of building in structural resistance to earthquakes is constantly evolving, producing surprising developments in technology. High-rise buildings in earthquake-prone areas are now designed to absorb, transmit, and otherwise survive at least moderate earthquakes (great earthquakes are another matter). This new technology has not yet been tested by numerous earthquakes, and only rarely by large ones. The 1995 earthquake in Kobe, Japan, was a very sobering event for engineers who claimed that Japan had developed state-of-the-art seismic resistance. Seismic engineers in the United States are still taking a wait-and-see attitude with regard to whether new buildings in San Francisco and Los Angeles, for instance, will survive "the big one."

Of course, no degree of built-in technology can ensure zero damage from a major event, just as all bets are off in a category 5 hurricane. However, the design and construction of earthquake resistance in single-family residences can be described as sound technology. This is because the forces that act on relatively small buildings are fairly well understood. It follows that the means to counter these forces is a matter of choice in the design and construction phases.

To be earthquake resistant, a structure must be built to resist violent horizontal (shear) strains, which tend to tear a building from its foundation. Taller buildings are subject to overturn. The taller the building, the greater this danger. To make a building earthquake resistant, one must strengthen all its weakest points and then anchor it to its foundation. Unreinforced rectangular walls, for instance, will deform into parallelograms. The presence of numerous windows increases structural weakness and makes reinforcement difficult. Compromises must be considered in the design phase. If a building is designed and built as a unit and then is well anchored to its foundation, it has a good chance of surviving a moderate earthquake intact. Earthquake-resistant construction is discussed in detail in chapter 9. Prospective homebuilders are encouraged to contact FEMA or the local building permitting authority for further information. Several excellent books are available on this subject (see appendix D, "Coastal Construction").

Fire!

The damage caused by the great Charleston earthquake and the memorable 1906 San Francisco earthquake was made many times worse by the ensuing fires. However, fires do not wait for earthquakes. Fire may strike at any time, for any number of reasons, and is far more likely to occur than an earthquake. Therefore it is logical, if not imperative, to mitigate this deadly hazard in the design of any structure. A nonfireproof building can be made more resistant to fire with locally available help and a little common sense. Information on the building codes that address fire safety can be found in appendix D under "Coastal Construction."

Historic Fires

Fire has never been known to play favorites; it strikes urban and rural structures alike. South Carolina history dating back to colonial days is filled with stories of tragedy in which frontier homes or barns ignited and were gone in minutes, often with loss of life. Though usually closer to help, urban areas have not been spared. When buildings are constructed close together, the danger that fires will spread is greatly increased. The city of Charleston has experienced many large fires. There were nine major fires, for example, in the hundred years between 1698 and 1796.

Charleston's greatest fire occurred in 1861. While Union ships blockaded the harbor and Charlestonians prepared for siege, a fire unrelated to the war broke out and burned from East Bay to Tradd and Rhett Streets, destroying more than a third of the city (fig. 7.5). The scars of this fire are visible in pictures taken during the Union occupation four years later.

What follows is, in our view, the best, most concise set of fire mitigation rules available, produced by the AEGIS Corporation of Nashville, Tennessee. The rules are based on a major research project designed to disclose the nature and causes of fires in the home and determine how residents can protect themselves from this primarily man-made hazard.

To prevent fires:

1. Install and maintain smoke detectors—they save lives.
2. Don't overload electric sockets. Replace or repair frayed cords.
3. Be vigilant about cigarettes and butts. Don't leave cigarettes unattended. Make sure butts are out. Never smoke in bed.
4. Protect children from fire hazards. Teach them about fire safety.

If a fire starts in your building:
1. *Get out!* Get out as quickly and directly as possible.
2. Check doors before opening them. If the door is hot, find another way out.
3. Stop, Drop, and Roll if your clothes ignite.
4. Once you're out, stay out!

Most fires start in kitchens. Follow these rules in the kitchen:

1. Never leave cooking food unattended.
2. Keep appliances clean. Built-up food and grease are flammable.
3. Don't store burnable or flammable items above the stove.
4. Wear tight sleeves when cooking. Avoid flowing clothes in the kitchen.
5. Turn off appliances when they are not in use.
6. Turn pot handles in so they do not hang over the edge of the stove.

Some of these rules may seem ridiculously simple, but don't be misled about their importance. They are based on statistics that tell what causes fires and what saves lives. Follow these commonsense rules and avoid potential tragedy.

No human habitation is completely fireproof. Even a fireproof building is filled with burnable materials, many of which give off toxic fumes. All buildings, regardless of the building materials, should be equipped with smoke alarms and fire extinguishers. Every person should be familiar with the exits, evacuation routes, and emergency plans for his or her home, temporary residence, or place of business. All building codes address these concerns (see appendix D). It remains up to you, the individual, to protect yourself, to educate your children and guests, and to take the steps necessary to prevent fires and fire-related deaths.

7.5. The great Charleston fire of 1861 destroyed the Cathedral of Saint John and Saint Finbar. Photograph courtesy of the South Carolina Historical Society.

Flooding

Barrier island flooding has already been addressed as one of the coastal hazards associated with hurricanes and storm surge (see chapters 2 and 3), but inland areas of the coastal zone are also subject to flooding (fig. 7.6). River floodplains dominate the landscape of the low country, and downstream flooding may occur from the runoff of inland storms as well as from the torrential rains associated with hurricanes. Fortunately, riverine flooding of larger rivers comes with adequate warning and moderate rates of rise in water levels, so evacuation is usually possible. Areas with greater relief and higher in elevation can experience flash flooding from even small streams.

Coastal flooding may be considered as two separate hazards according to the different problems it presents. Rising water from swollen coastal rivers is a hazard to lives and property. Standing water from locally heavy rainfall is common in coastal areas and may be little more than a nuisance. There is a considerable difference in the mitigation strategies for these two levels of flooding and in the long-term likelihood of success of mitigation efforts.

One Historic Flood: A Case in Point

A terrible flood inundated the city of Charleston in the year 1796. Little was recorded about the loss of life and the damage to colonial homes, but we do know that water flowed waist-deep across the peninsula. Charlestonians and residents of the sea islands had learned a thing or two about hurricanes and the floods that came with them by then. But this calamity was different. The flood came without warning; even more unusual, it came without a hurricane. In typical Charlestonian understatement it was called the "freshet of '96."

The "freshet" was the result of a storm or series of storms that essentially bypassed the coastal zone and poured water over the piedmont region of South Carolina. The floodwaters came to Charleston not from the sea but via the Santee River, which drains the piedmont. The Santee River flows nowhere near the Holy City, of course. But the freshet of '96 put so much water into the swollen Santee that it overflowed its banks. The floodwaters were diverted into the Cooper River, scouring a breach in the divide between the rivers in the process.

The freshet of '96 is significant today because the location of the scoured breach was selected as the most logical location of the Santee-Cooper Diversion Project, which diverted Santee River waters into the Cooper River, a hapless venture intended to save the $5 million per year it cost to dredge Charles-

ton Harbor. Engineers thought that the increased river flow would sweep the harbor free of sediment deposition. The Santee, however, delivered only more sediment—much more—and none of the expected flushing action. The yearly price tag immediately rose to $50 million. The 1980s brought us the Santee Rediversion Project. Lord, save us from the things we do in the name of flood control!

Sea level rise also plays a role in river flooding. In chapter 2 we saw how the rising sea level drives barrier island migration. But the rising sea level influences more than just the shoreline. River drainage systems and floodplains are also flooded, along with land area. They don't call this the low country for nothing! And more flooding of the low areas

7.6. Nuisance flooding of low-lying areas is a problem that property buyers from inland areas may not anticipate. Coastal storms often generate heavy rainfall that overwhelms storm drainage systems.

means less area to handle runoff of rainwater from major storms, which will increase the potential for damage from flooding. The coastal zone is a hazardous place.

*Mitigation: God Willin' and
the River Don't Rise*

Little has changed with regard to effective flood control since 1796. We have dredged and filled along many of our riverbanks, but when water levels rise, for whatever reason, the land is still inundated. Our municipalities have the responsibility to mitigate such floods. This leads to sometimes grandiose and always expensive flood control measures. Charleston's newest project, a multimillion-dollar pumping and dewatering project, is under construction. It is, of course, intended to reduce flooding in parts of the peninsula, and it may be completed in time to compensate for the sea level rise of the 1990s. But South Carolinians are not alone. Many coastal cities of the world are at or below sea level—New Orleans, Venice, London, Amsterdam, and Singapore, to name a few—and each has its own plan to mitigate flooding: locks, canals, dams, dikes, and pumping out water that once flowed downhill to the sea. Flooding is a problem we humans haven't heard the last of.

The greatest problem with respect to flooding is that modern development is spreading out onto the floodplains at an alarming rate. Our poor understanding of flood history and our tendency to ignore what the land tells us are placing an increasing number of people in the flood hazard zone, not to mention millions of dollars in property investments—raising the price we will all pay in the next flood, large or small. In addition, flood "control" robs fertile farmlands of their major source of new organic material. For eons, river floods provided organic-rich sediment, layer after layer. With this natural source of organic matter cut off (in the name of progress!), we increasingly have to use chemical fertilizers to grow our crops.

Minor Mitigation: Cover Your Assets

The solution to flooding, first and foremost, is to avoid the hazard by avoiding the floodplain. Buildings already located in harm's way should be floodproofed to reduce future losses (see chapter 8). If your land has standing water after a rain, flooding can be less a hazard and more just an aggravation. Water from seasonal rains often fills low areas such as dune swales. Nature has the audacity to do this even when the low area in question is your yard, the road you take to work, or even your house or business. Standing water is a nuisance. It can ruin your yard or foundation, it provides a breeding place for mosquitoes, and it is likely to be polluted.

If your property is in a low area, there is one technique that usually works to mitigate nuisance flooding: elevate. If your house gets flooded, elevate it. If your yard gets flooded, elevate that. On a small scale, raising the elevation of your yard by trucking in sand is usually quite effective. (This may really aggravate your neighbors if they don't also elevate, but that's another hazard.) Permits to do this work are usually not a problem. Sand is usually inexpensive and it drains well, which is a requirement. This concludes the list of hazards associated with living with the South Carolina coast. The good news is that South Carolina's coastal zone is immune to tidal waves, landslides, volcanoes, icebergs, and a host of other hazards common in other coastal areas of the world. The bad news is that a hurricane can be just as devastating to individuals and property as a tsunami.

At least our coastal hazards can be mitigated. In our rush to occupy coastal areas it is increasingly imperative that we address these hazards to reduce their toll on lives and properties. We rely on our civic leaders for flood control and building codes. But in the long run, the mitigation of coastal hazards rests primarily with the individual.

All of which brings us back full circle to the point of this book. Let it be stated again here. The boiled-down, crystallized, core point of *Living with the South Carolina Coast* is *mitigate!* Don't wait. Don't point the finger. Begin today. Take the necessary steps to prepare for and protect yourself and your property from the inherent dangers of life in the coastal zone. Choose a good building site, check your smoke alarms, and know your evacuation route. The mitigation of coastal hazards begins at home, your home. In these days of insurance from the cradle to the grave, of lawsuits over spilled coffee, a hefty insurance check will avail you nothing if you sleep with the fishes.

In reading this book you may conclude that its authors are at cross purposes. On the one hand, we point out that building on the coast is risky. On the other hand, we provide you with a guide to evaluating the risks, and in this chapter we advise how best to buy or build a house in the coastal zone.

This apparent contradiction is more rational than it might seem. For those who will heed the warning, we describe the risks of owning shorefront property. But we realize that coastal development will continue. Some individuals will always be willing to gamble with their fortunes to be near the shore. For those who elect to play this game of real estate roulette, we offer some advice on improving the odds, on reducing (not eliminating) the risks. We do not recommend, however, that you play the game!

Can We Learn from Past Experience?

Although the memory of Hugo is fading, coastal property owners in South Carolina would be foolish to ignore the problems the hurricane brought to light. Similarly, the lessons learned from hurricanes in other areas (e.g., Andrew, Gilbert, and Iniki) must be heeded. Why, for example, did Habitat for Humanity houses built by amateurs stand up to Andrew while houses built by professionals blew apart? What have posthurricane damage inspections revealed, and what do teams of mitigation experts recommend?

If you want to learn more about construction near the beach, several sources containing detailed information on coastal construction are listed in appendix D, under "Coastal Construction," along with additional references.

Coastal Realty versus Coastal Reality

Coastal property is not like inland property. Do not approach it as if you were buying a lot in a developed woodland of the Carolina piedmont or a subdivided farm field in the coastal plain. The previous chapters illustrate that the shores of South Carolina, especially the barrier islands, are composed of variable environments and are subject to nature's most powerful and persistent forces. The reality of the coast is its dynamic character. Property lines are an artificial grid superimposed on this dynamism. If you choose to place yourself or others in this zone, prudence is in order.

A quick glance at the architecture of the structures on our coast provides convincing evidence that the reality of coastal processes was rarely considered in their construction. Instead, the sea view and aesthetics seem to have been the primary considerations. Except for meeting minimal building code requirements, no further thought seems to have been given to the safety of many of these buildings. The failure to follow a few basic architectural guidelines that recognize this reality will have disastrous results in the next major storm.

Life's important decisions should be based on a thorough evaluation of the facts. Few of us buy large items, choose a career, or take legal, financial, or medical actions without first evaluating the facts and seeking advice. In the case of coastal property, two general aspects should be evaluated: site safety and the integrity of the structure relative to the forces to which it will be subjected.

We have so far presented a guide to the various environments on the South Carolina open-ocean shoreline (chapter 5) and have provided risk maps that help to evaluate hazards at specific sites (chapter 6). This chapter focuses on the structures that are built on the shore—houses, cottages, and condominiums.

The Structure: Concept of Balanced Risk

Any structure built within the constraints of economy and environment can fail. The objective of a building design is to create a structure that is both economically feasible and functionally reliable. That is, a house must be affordable and should have a reasonable life expectancy. In order to obtain such a house, a balance must be achieved among financial, structural, environmental, and other special conditions. Most of these conditions are heightened on the coast: property values are higher, there is a greater desire for aesthetics, the environment is more sensitive, the likelihood of storms is increased, and there is greater pressure to develop as more and more people want to move into the coastal zone.

The individual who builds or buys a home in an exposed area should fully comprehend the risks involved and the chance of harm to home or family. The risks should then be weighed against the benefits to be derived from the residence. Similarly, the developer who is putting up a motel should weigh the possibility of destruction and death during a hurricane against the money and other advantages to be gained from such a building. Then and only then should construction proceed. For both the homeowner and the developer, proper construction and location reduce the risks involved. The concept of balanced risk should take into account six fundamental considerations:

1. Construction must be economically feasible.
2. Because construction must be economically feasible, ultimate and total safety is not obtainable for most homeowners on the coast.
3. A coastal structure is exposed to high winds, waves, and flooding and should be stronger than a structure built inland.
4. A building with a planned long life, such as a year-round residence, should be stronger than a building with a planned short life, such as a mobile home.
5. A building with high occupancy, such as an apartment building, should be safer than a building with low occupancy, such as a single-family dwelling.

6. A building that houses elderly or sick people should be safer than a building housing able-bodied people.

Structures can be designed and built to resist all but the largest storms and still be reasonably economical. Structural engineering involves the design and construction of buildings to withstand the forces of nature. It is based on a knowledge of the forces to which the structures will be subjected and an understanding of the strength of building materials. The effectiveness of structural engineering design was reflected in the aftermath of Cyclone Tracy, which struck Darwin, Australia, in 1974: 70 percent of the housing that was not based on structural engineering principles was destroyed and 20 percent was seriously damaged—only 10 percent of the houses weathered the storm. In contrast, more than 70 percent of the large, structurally engineered commercial, government, and industrial buildings came through with little or no damage, and less than 5 percent of such structures were destroyed. Because housing accounts for more than half of the capital cost of the buildings in Queensland, Australia, the state government there established a building code requiring standardized structural engineering for houses in hurricane-prone areas. This improvement has been achieved with little increase in construction and design costs. In contrast, South Carolina has yet to adopt stronger code requirements in the post-Hugo era, and new construction will

probably suffer the same fate as pre-Hugo buildings in the next big storm.

Can We Rely on Building Codes?

Peter Sparks, a civil engineer, noted in the wake of Hugo that wind damage often was initiated by the loss of the roof structure, which led to progressive building collapse (appendix D, ref. 39). Similarly, the failure of doors and windows exposed to high winds led to additional building damage, or even complete destruction. Dr. Sparks and other investigators (appendix D, ref. 160) point out that since 1972 South Carolina has permitted, but not required, local governments to use the (Southern) Standard Building Code. Unfortunately, this code lacks some of the specifications required for hurricane-resistant construction. In particular, the Standard Building Code's framers did not use the national loading standard for wind loads! The effects of such special considerations as wind gusts, higher suction created at wall corners and roof edges, and exposure to open-ocean areas are not included in the code. Furthermore, code enforcement was difficult prior to Hugo for a number of reasons. Hugo taught owners, buyers, lenders, insurers, and community officials that the mere existence of a building code is no guarantee of building safety or stormworthiness. Dr. Sparks concluded that even if strict construction requirements are imposed, it will be many years

Wind

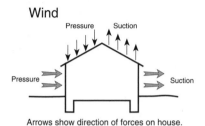

Arrows show direction of forces on house.

Drop in barometric pressure

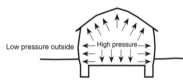

The passing eye of the storm creates different pressures inside and out; high pressure inside attempts to burst house open.

Waves

House is damaged by the force of the waves.

High water

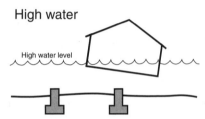

Unanchored house floats off its foundation

8.1. Forces to be reckoned with at the shoreline.

before each community reaches the point where the majority of its buildings can resist hurricanes. Nevertheless, we must heed Herbert Saffir's rule: "Coastal construction must conform to higher standards than inland construction" (appendix D, ref. 161).

Coastal Forces: Design Requirements

Northeasters can be devastating, but hurricanes are the most destructive forces that must be reckoned with on the coast (fig. 8.1). Coastal storms are discussed in detail in chap-

ter 3, but let us restate some of the most important points and add a bit more information here for emphasis.

Hurricane Winds

Hurricane winds can be evaluated in terms of the pressure they exert. A 100-miles-per-hour wind exerts a pressure or force of about 40 pounds per square foot on a flat surface. The pressure varies with the square of the velocity. Thus, a wind with a velocity of 190 miles per hour exerts a force of 144 pounds per square

foot. This force is modified by several factors which must be considered in designing a building. For instance, the pressure on a round surface, such as a sphere or cylinder, is less than the pressure on a flat surface. Also, winds increase with the height above the ground, so a tall structure is subject to greater pressure than a low structure.

A house or building designed for inland areas is built primarily to resist vertical, and mostly downward, loads. Builders assume that the foundation and framing must support the load of the walls, floor, and roof, and relatively insignificant horizontal wind forces. A well-built house in a hurricane-prone area, however, must be constructed to withstand a variety of strong wind *and wave* forces that may come from any direction. Although many people think that wind damage is caused by uniform horizontal pressures (lateral loads), most of the damage is caused by uplift (vertical), suctional (pressure-differential), and torsional (twisting) forces. High horizontal pressure on the windward side is accompanied by suction on the leeward side. The roof is subject to downward pressure and, more important, to uplift. Often roofs are sucked up by the uplift drag of the wind. Usually houses fail because the devices that tie the parts of the structure together fail. All structural members (beams, rafters, and columns) should be fastened together on the assumption that about 25 percent of the vertical load on each member may be a force coming from any direction (sideways or upward). Such structural integrity is also important if it is likely that the building may be moved to avoid destruction by shoreline retreat. A fanciful way of understanding structural integrity is that you should be able to pick up a house (after removing its furniture, of course), turn it upside down, and shake it without it falling apart.

Storm Surge

Storm surge is a rise in sea level above the normal water level during a storm. During hurricanes the coastal zone is inundated by storm surge and accompanying storm waves, and these cause most of the property damage and loss of life.

Often, the pressure of the wind backs water into streams or estuaries already swollen from the exceptional rainfall brought by the hurricane. Water is piled into the bays between islands and the mainland by the offshore storm. In some cases the direction of flooding may be from the bay side of the island. This flooding is particularly dangerous when the wind pressure keeps the tide from running out of inlets, so that the next normal high tide pushes the accumulated waters back, and higher still.

People who have cleaned the mud and debris out of a house subjected to flooding retain vivid memories of its effects. Flooding can cause an unanchored house to float off its foundation and come to rest against another house, severely damaging both. But even if the house itself is left structurally intact, flooding may destroy its contents.

Proper coastal development takes into account the expected level and frequency of storm surge for the area. In general, building standards require that the first habitable level of a dwelling be above the 100-year flood level. At this level, a building has a 1 percent probability of being flooded in any given year.

Hurricane Waves

Hurricane waves can cause severe damage not only through forcing floodwaters onshore but also by throwing boats, barges, piers, houses, and other floating debris inland against standing structures. In addition, waves can destroy coastal structures by scouring away the underlying sand, causing them to collapse. It is possible to design buildings to survive crashing storm surf. Many lighthouses (e.g., the Morris Island light) have survived hurricane waves. But in the balanced-risk equation, it usually isn't economically feasible to build ordinary cottages to withstand the more powerful forces. On the other hand, cottages can be made considerably more stormworthy by following the suggestions in the following sections.

The force of a wave may be understood by considering that a cubic yard of water weighs more than three-fourths of a ton. A breaking wave moving shoreward at a speed of 30 or 40 miles per hour can be one of the most destructive elements of a hurricane. A 10-foot wave can exert more than a thousand pounds of pressure per square foot, and wave pressures as high as 12,700 pounds per square foot have been recorded. Figure 8.2 illustrates some of the actions that a homeowner can take to deal with the forces just described.

Construction Type: Lessons from Gilbert and Hugo

Building materials and construction methods had a profound effect on the relative storm damage to buildings in Hurricane Hugo's path compared with damage caused by Hurricane Gilbert (appendix D, ref. 132), the most powerful storm to strike the Western Hemisphere thus far this century. Gilbert's central barometric pressure of 885 millibars is the lowest ever recorded in the Caribbean. After mauling Jamaica, Gilbert made landfall as a category 5 hurricane over the Yucatán Peninsula of Mexico near Cozumel and moved west-northwest across the peninsula.

Single-family structures on the Yucatán coast are built primarily of reinforced concrete. Construction lumber is extremely expensive there, there is an essentially limitless supply of limestone for making concrete, and most Mexican engineers and architects are trained in Mexico City, where reinforced concrete construction is used extensively to mitigate earthquake damage. Also, Yucatán homes are not elevated; many are built right into the frontal line of dunes.

A number of building materials are used on the South Carolina coast. Single-family homes are typically built of wood, cinder block, or a combination of wood, brick, and cinder block. Newer structures are elevated up to about 15 feet on wooden or concrete pilings in order to comply with flood insurance regulations. Older structures are elevated on wooden pilings or cinder-block footings. Generally speaking, the older the structure, the less elevated it is likely to be.

The use of different construction materials produced substantial differences in storm damage in the two areas. When buildings along the Yucatán oceanfront collapsed, the reinforcing rods in the concrete held the large pieces in place. Their rubble formed a temporary seawall-like protection for the more landward buildings, reducing inland damage. When South Carolina cinder-block houses failed, they broke into small pieces that were tossed about and damaged other structures. In addition, inadequately elevated wooden houses were knocked off their foundations and carried some distance from their original locations. In many places, oceanfront houses were pushed back into the third or fourth row of development, in the process colliding with other houses, trees, utility poles, and automobiles. At least five houses were completely removed

from Pawleys Island; three were later found in the back-barrier marsh nearly a mile from their original location.

Reinforced concrete construction is not typically thought of as aesthetically or economically appealing in the United States, although it is common in some states such as Florida and Hawaii, and in Puerto Rico.

House Selection

Some types of houses are better at the shore than others, and an awareness of the differences will help you make a better selection,

whether you are building a new house or buying an existing one.

Worst of all are unreinforced masonry houses, whether they be brick, concrete block, hollow clay-tile, or brick veneer, because they cannot withstand the lateral forces of winds and waves and the settling of the foundation. Extraordinary reinforcing will alleviate some of the inherent weaknesses of unit masonry if done properly, but generally this type of construction is inappropriate for coastal regions.

8.2. Modes of structural failure and how to deal with them.

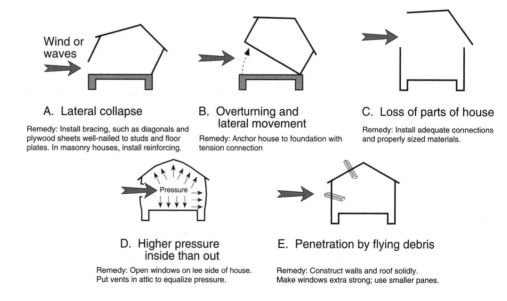

A. Lateral collapse
Remedy: Install bracing, such as diagonals and plywood sheets well-nailed to studs and floor plates. In masonry houses, install reinforcing.

B. Overturning and lateral movement
Remedy: Anchor house to foundation with tension connection

C. Loss of parts of house
Remedy: Install adequate connections and properly sized materials.

D. Higher pressure inside than out
Remedy: Open windows on lee side of house. Put vents in attic to equalize pressure.

E. Penetration by flying debris
Remedy: Construct walls and roof solidly. Make windows extra strong; use smaller panes.

Reinforced concrete and steel frames are excellent but are rarely used in small residential structures. In Puerto Rico, where hurricanes are more a way of life than in South Carolina, residences are typically built of cast-in-place concrete or concrete masonry units with reinforced concrete columns and perimeter beams. The roof slabs are approximately 4 to 5 inches thick. The design is governed by seismic codes like those used in Mexico, and this class of structure performed extremely well in Hurricane Hugo (appendix D, ref. 40).

It is hard to beat a wood-frame house that is properly braced and anchored and has well-connected members. A well-built wood house will often hold together as a unit—even if moved off its foundation—when other types disintegrate. Although all the structural types noted above are found in the coastal zone, newer structures tend to be of the elevated wood-frame type.

The Exterior Envelope

The term *building envelope* refers to the entire system by which the building resists wind penetration. A breach in the envelope occurs when an exterior enclosure fails, such as when the garage door or a window is open. During a strong wind, such an opening allows the pressure to build up inside, and roof uplift or wall suction may occur, leading to failure of the entire system. The most susceptible parts of the house are the windows, garage doors, and double entry doors.

All doors in a home should be certified by the seller as to their strength under a given design wind load, especially garage doors and double entry doors. Their strength must be adequate to prevent damage from projectiles or flying objects such as tree limbs. Locking systems such as wind-resistant latches and dead

8.3 Where to strengthen doors. *Source: Against the Wind* (appendix D, ref. 21).

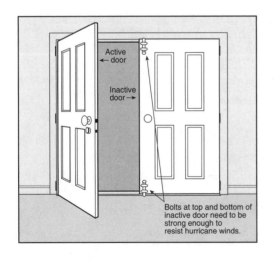

Active ← door

Inactive door →

Bolts at top and bottom of inactive door need to be strong enough to resist hurricane winds.

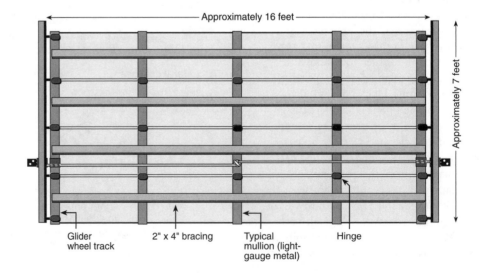

Approximately 16 feet

Approximately 7 feet

Glider wheel track

2" x 4" bracing

Typical mullion (light-gauge metal)

Hinge

bolts can be used to reinforce existing doors. The dead bolt will act as an additional rigid connection to the house frame. If the house has a double entry door, one door should be fixed at the top and bottom with pins or bolts (fig. 8.3). The original pins are usually not strong enough to resist heavy wind forces. Homeowners should consider installing heavy-duty bolts (your local hardware store will be able to advise you as to the strength of the bolts or pins).

Garage doors pose another risk to the building envelope. They often fail because of inadequate thickness, and they tend to bend when subjected to strong winds. Double-wide or two-car garage doors are especially susceptible to high winds. New and existing garage doors can be fixed to alleviate this problem. When purchasing a new garage door, the manufacturer's certification of its strength should be verified for the adequacy of the system. Retrofitting existing garage doors can be done by installing horizontal girts on each panel (fig. 8.3). A less permanent strategy to reinforce your garage door during a wind storm is to back your car up against the inside of the door, providing extra support against bending. To prevent the garage door from falling off its tracks during high winds, strengthen the track supports and glider tracks. The rotation of the door along its edges can be reduced by chaining the door pin to the glider track connections.

All windows—including skylights, sliding glass doors, and french doors—must be protected from projectiles that could penetrate the building envelope. During a storm, such missiles may be branches, roof pieces, lawn furniture, or anything else that can be picked up by the wind. Windows can be protected by using storm shutters, but it is also important that the windows be strong enough to stand up to high winds. In commercial or office buildings where shutters are impractical, windows can be reglazed. Reglazing is a method of strengthening the windows that replaces normal glass with tempered laminated glass. Finally, all windows and doors should be anchored to the wall frame to prevent them from being pulled out of the building.

Keeping Dry: Pole or "Stilt" Houses

In coastal regions subject to flooding by waves or storm surge, the best and most common method of minimizing damage is to raise the lowest floor of a residence above the expected water level. The first habitable floor of a home must be above the 100-year storm-surge level (plus calculated wave height) to comply with local regulations. Most modern structures built in flood zones are constructed on piling that is well anchored in the subsoil. Elevating the structure by building a mound is not a suitable strategy in the coastal zone because mounded soil is easily eroded. Construction on piles or columns is required within the V-zone.

Current building design criteria for pole house construction under the National Flood Insurance Program (NFIP) are outlined in the coastal manual and the book *Elevated Residential Structures* (appendix D, ref. 173). Regardless of insurance requirements, pole-type construction with deeply embedded poles is best in areas where waves and storm surge will erode foundation material. Materials used in pole construction include piles, posts, and piers.

Piles are long, slender columns of wood, steel, or concrete driven into the earth to a depth sufficient to support the vertical load of the house and withstand the horizontal forces of flowing water, wind, and waterborne debris. Pile construction is especially suitable in areas where scouring (soil "washing out" from under the foundation of a house) is a problem.

Posts are usually made of wood (if steel, they are called columns). Unlike piles, posts are not driven into the ground, but instead are placed in a predug hole which may or may not have a concrete pad at the bottom (fig. 8.4). Posts may be held in place by backfilling and tamping earth or by pouring concrete into the hole after the post is in place. Posts are more readily aligned than driven piles, and are therefore better to use if the poles must extend to the roof. In general, treated wood is the cheapest and most common material for both posts and piles.

Piers are vertical supports, thicker than piles or posts and usually made of reinforced concrete or reinforced masonry (concrete blocks or bricks). They are set on footings and extend to the underside of the floor frame. This type of foundation was very vulnerable to water erosion during Hugo; pier footings toppled and piers buckled, leading to building damage or collapse (appendix D, ref. 39).

Pole construction can be of two types. The poles can be cut off at the first floor to support the platform that serves as the dwelling floor. In this case, piles, posts, or piers can be used. Or the poles can be extended to the roof and rigidly tied into both the floor and the roof, thus becoming major framing members for the structure and providing better anchorage to the house as a whole (figs. 8.5, 8.6). Sometimes both full-height and floor-height poles are used, with the shorter poles restricted to supporting the floor inside the house (fig. 8.6).

On sites where the foundation material can be eroded by waves or winds, the poles should be deeply embedded and solidly anchored, either by driving piles, by jetting them, or by drilling deep holes for posts and putting in a concrete pad at the bottom of each hole. Where the embedment is shallow, a concrete collar around the poles improves the anchorage (fig. 8.4). The choice depends on the soil conditions. Piles are more difficult than posts to align to match the house frame; posts can be positioned in the holes before backfilling. Inadequate piling depths, improper piling-to-floor connections, and inadequate pile bracing all contribute to structural failure when storm waves liquefy and erode the sand that supports them.

If posts are used instead of pile, the posts should extend 6 to 8 feet into the ground to provide anchorage. The lower end of each post should rest on a concrete pad, which spreads the load on the soil over a greater area to prevent settling. If the soil is sandy or the embedment less than about 6 feet, it is best to tie the post down to the footing with straps or other anchoring devices. Driven or jetted piles should extend to a depth of 10 feet or more.

The floor and roof should be securely connected to the poles with bolts or other fasteners. If the floor rests on poles that do not extend to the roof, attachment is even more critical. A system of metal straps is often used. Another method is to attach beams to piles with at least two bolts of adequate size. Unfortunately, builders sometimes simply attach the floor beams to a notched pole with one or two bolts. Hurricanes have proven this method inadequate. During the next hurricane on the South Carolina coast, many houses will be destroyed because of inadequate attachment.

Local building codes may specify the size, quality, and spacing of the piles, ties, and brac-

8.4. Shallow and deep supports for poles and posts. Supplied by the Southern Pine Association.

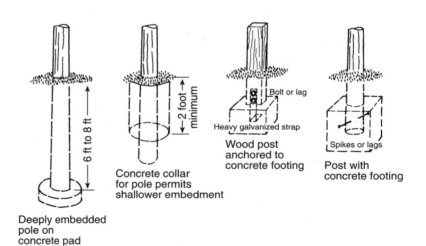

ing, as well as the methods of fastening the structure to them. Building codes often are minimal requirements, however, and building inspectors are usually amenable to allowing designs that are equally or more effective.

The space under an elevated house, pole-type or otherwise, must be kept free of obstructions in order to minimize the impact of waves and floating debris. If the space is enclosed, the enclosing walls should be designed so that they can break away or fall under flood loads but also either remain attached to the house or be heavy enough to sink. In other words, you don't want the walls to float away and add to the waterborne debris problem. Alternatively, enclosing walls can be designed to be swung up, out of the path of floodwaters, or built with louvers that allow the water to pass through (louvered walls are subject to damage from floating debris, however). The convenience of closing in the ground floor for a garage or extra bedroom may be costly because it may violate local ordinances and may result in increased insurance premiums, and it may actually contribute to the loss of the house in a hurricane.

An Existing House: What to Look For, Where to Improve

If, instead of building a new house, you are selecting a house already built in an area subject to flooding and high winds, consider (1) where

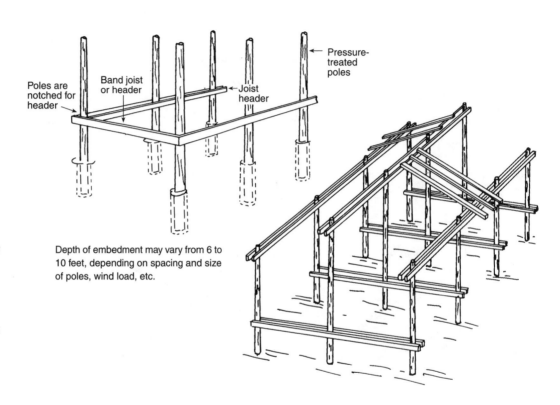

8.5. Framing system for an elevated house. Supplied by the Southern Pine Association.

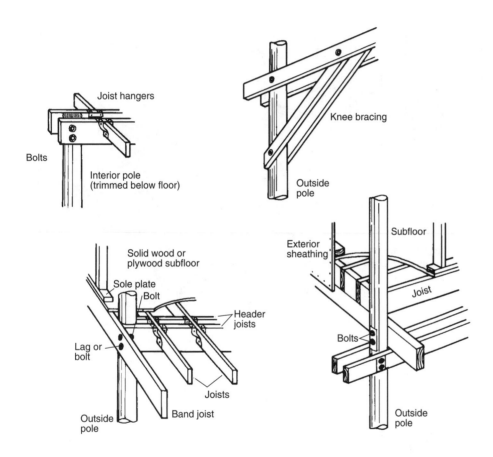

Joist hangers

Bolts

Interior pole
(trimmed below floor)

Knee bracing

Outside
pole

Exterior
sheathing

Subfloor

Joist

Bolts

Outside
pole

Solid wood or
plywood subfloor

Sole plate

Bolt

Header
joists

Lag or
bolt

Joists

Outside
pole

Band joist

8.6. Tying floors to poles. Supplied by the Southern Pine Association.

the house is located, (2) how well the house is built, and (3) how the house can be improved.

Location

Evaluate the site of an existing house using the same principles you would use to evaluate a building site. The elevation of the house, frequency of high water, escape route, and how well the lot drains should be emphasized, but you should go through the complete 13-point site-safety checklist given in chapter 5.

You can modify the house after you have purchased it, but you can't prevent hurricanes or northeasters. First, stop and consider: Do the pleasures and benefits of this location balance the risks and disadvantages? If not, look elsewhere for a home; if yes, then proceed to evaluate the house itself.

How Well Built Is the House?

In general, the principles used to evaluate an existing house are the same as those used when building a new one (see appendix D, "Coastal Construction").

Before you thoroughly inspect the house in which you are interested, look closely at the adjacent homes. If they are poorly built, they may float over against your house and damage it in a flood. You may even want to consider the type of people you will have as neighbors: Will they "clear the decks" in preparation for a storm, or will they leave items in the yard to become windborne missiles? The house itself should be inspected for the following features.

The house should be well anchored to the ground, and the walls well anchored to the foundation. If the house is simply resting on blocks, rising water may cause it to float off its foundation and come to rest against your neighbor's house or out in the middle of the street. If the house is well built and well braced internally, it may be possible to move it back to its proper location, but chances are great that the house will be too damaged to be habitable.

If the house is on piles, posts, or poles, check to see if the floor beams are adequately bolted to them. If it rests on piers, crawl under the house, if space permits, to see if the floor beams are securely connected to the foundation. If the floor system rests unanchored on piers, do not buy the house.

It is difficult to discern whether a house built on a concrete slab is properly bolted to the slab because the inside and outside walls hide the bolts. If you can locate the builder, ask if such bolting was done. Better yet, if you can get assurance that construction of the house complied with the provisions of a building code serving the needs of that particular region, you can be reasonably sure that all parts of the house are well anchored: the foundation to the ground, the floor to the foundation, the walls to the floor, and the roof to the walls (figs. 8.7–8.10).

Be aware that many builders, carpenters,

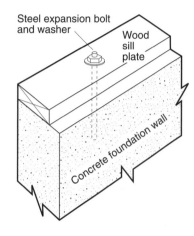

8.7. Additional connection of a wood-frame building to its foundation. Temporarily remove the wall covering enough to add half-inch or larger steel expansion bolts to gain additional anchorage. *Sources:* Figures 8.7–8.12 are modified from *Coastal Design: A Guide for Builders, Planners, and Homeowners; Building Performance: Hurricane Andrew in Florida; Building Performance: Hurricane Iniki in Hawaii;* and *Coastal Construction Manual* (appendix D, refs. 172, 183, 184, and 186, respectively).

and building inspectors accustomed to traditional construction are apt to regard metal connectors, collar beams, and other such devices as newfangled and unnecessary. If consulted, they may assure you that a house is as solid as a rock when in fact it is far from it. Nevertheless, it is wise to consult the builder or knowledgeable neighbors when possible.

The roof should be well anchored to the walls to prevent uplifting and separation from the walls. Visit the attic to see if such anchoring exists. Simple toe-nailing (nailing at an angle) is not adequate; metal fasteners are needed. Depending on the type of construction and the amount of insulation laid on the floor of the attic, these may or may not be easy to see. If roof trusses or braced rafters were used, it should be easy to see whether the various members, such as the diagonals, are well fastened together. Again, simple toe-nailing will not suffice. Some builders, unfortunately, nail parts of a roof truss just enough to hold it together to get it into place. A collar beam or gusset at the peak of the roof (fig. 8.11) provides some assurance of good construction.

Good-quality roofing material should be well anchored to the sheathing. A poor roof covering will be destroyed by hurricane-force winds, allowing rain to enter the house and damage the ceilings, walls, and contents. Galvanized nails (two per shingle) should be used to connect wood shingles and shakes to wood sheathing, and they should be long enough to penetrate through the sheathing (fig. 8.11). Threaded nails should be used for plywood sheathing. For roof slopes that rise 1 foot for every 3 feet or more of horizontal distance, exposure of the shingle should be about one-fourth of its length (4 inches for a 16-inch shingle). If shakes (thicker and longer than shingles) were used, less than one-third of their length should be exposed.

Steel expansion bolt and washer

Wood sill plate

Concrete foundation wall

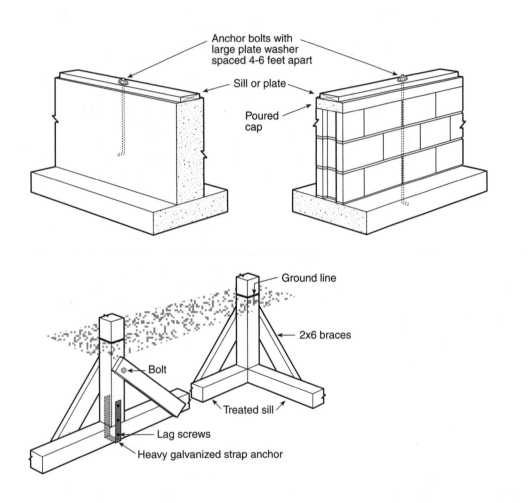

Anchor bolts with large plate washer spaced 4-6 feet apart

Sill or plate

Poured cap

Ground line

2x6 braces

Bolt

Treated sill

Lag screws

Heavy galvanized strap anchor

8.8. Foundation anchorage. (Top) Anchored sill for shallow embedment. (Bottom) Anchoring sill or plate to foundation.

In hurricane-prone areas, asphalt shingles should be exposed somewhat less than usual. A mastic or seal-tab type, or an interlocking shingle of heavy grade should be used, along with a roof underlay of asphalt-saturated felt and galvanized roofing nails or approved staples (six for each three-tab strip).

Asphalt shingles are not the best roofing material to use in high-wind areas. Wind engineers, structural engineers, builders, and insurers are working together to develop new roof technology. The battle cry is "Reinvent roofing for coastal areas." Specifically, the research involves a search for better roof-covering materials. Perhaps most important is to replace asphalt shingles with some sort of continuous

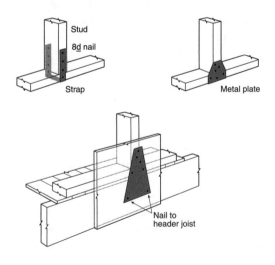

Stud

8d nail

Strap

Metal plate

Nail to header joist

8.9. Stud-to-floor, plate-to-floor framing methods.

waterproof membrane that wind cannot lift up and tear off.

With regard to framing, the fundamental rule to remember is that all the structural elements should be fastened together and anchored to the ground in such a manner as to resist all forces, regardless of which direction they come from. This prevents overturning, floating off, racking, or disintegration.

The shape of the house is important. A hipped roof, which slopes in four directions, is better able to resist high winds than a gabled roof, which slopes in two directions. This was found to be true in Hurricane Camille (1969) in Mississippi, in Cyclone Tracy (1974) in Australia, and in more recent Atlantic coast hurricanes. The reason is twofold: a hipped roof offers a smaller shape for the wind to blow against, and it is better braced in all directions.

Also note the horizontal cross section of the house (the shape of the house as viewed from above). The pressure exerted by a wind on a round or elliptical shape is about 60 percent of that exerted on a square or rectangular shape; the pressure exerted on a hexagonal or octagonal cross section is about 80 percent of that exerted on a square or rectangular cross section.

The design of a house or building in a coastal area should minimize structural discontinuities and irregularities. A house should have a minimum of nooks and crannies and offsets on the exterior, because damage to a structure tends to concentrate at these points. Many of the newer beach houses along the

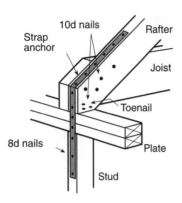
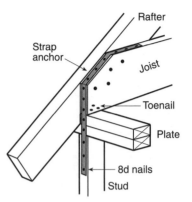
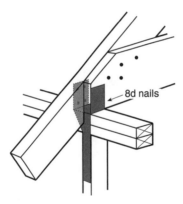
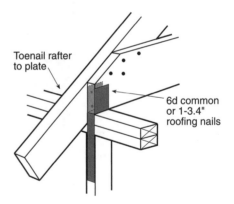

8.10. Roof-to-wall connectors. (Top) Metal strip connectors: (left) rafter to stud; (right) joist to stud. (Bottom left) Double-member metal plate connector, in this case with the joist to the right of the rafter. (Bottom right) A single-member metal plate connector.

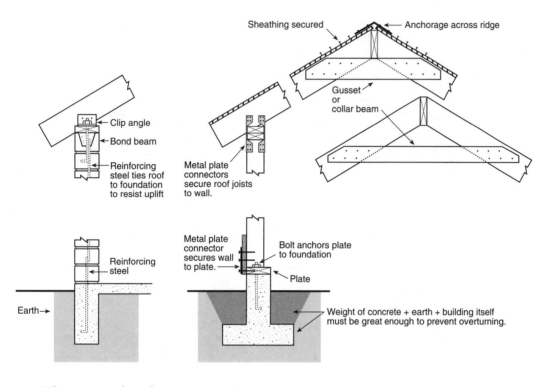

Sheathing secured

Anchorage across ridge

Gusset
or
collar beam

Clip angle

Bond beam

Reinforcing
steel ties roof
to foundation
to resist uplift

Metal plate
connectors
secure roof joists
to wall.

Reinforcing
steel

Metal plate
connector
secures wall
to plate.

Bolt anchors plate
to foundation

Plate

Earth→

Weight of concrete + earth + building itself
must be great enough to prevent overturning.

8.11. Where to strengthen a house.

U.S. coast are of a highly angular design with just such nooks and crannies. Award-winning architecture will be a storm loser if the design has not incorporated the technology for maximizing structural integrity with respect to storm forces. When irregularities are absent, the house reacts to storm winds as a complete unit.

Brick, concrete-block, and masonry-wall houses should be adequately reinforced. This reinforcement is hidden from view (fig. 8.11). Building codes applicable to high-wind areas often specify the type of mortar, reinforcing, and anchoring to be used in construction. If you can get assurance that the house was built in compliance with a building code designed for such an area, consider buying it. At all costs avoid unreinforced masonry houses.

Even if reinforced, masonry structures are *not* recommended in V-zones.

A poured-concrete bond beam at the top of the wall just under the roof is one indication that the house is well built (fig. 8.12). Most bond beams are formed by putting in reinforcing and pouring concrete in U-shaped concrete blocks. From the outside, however, you can't distinguish these U-shaped blocks from ordinary ones, and therefore you can't be certain that a bond beam exists. The vertical reinforcing should penetrate the bond beam.

Some architects and builders use a stacked bond (one block directly above another) rather than overlapped or staggered blocks because they believe it looks better. The stacked bond is definitely weaker than the overlapped or staggered blocks. Unless you have proof that the walls are adequately reinforced to overcome this lack of strength, you should avoid this type of construction. Remember: some masonry-walled buildings completely collapsed in Hugo, resembling the flattened buildings associated with earthquakes.

In past hurricanes, the brick veneer of many houses separated from the wood frame, even when the houses remained standing. The asbestos-type outer-wall panels used on many houses in Darwin, Australia, were brittle and broke up under the impact of windborne debris in Cyclone Tracy. Gypsum-board cladding (covered with insulation and stucco) performed very poorly in Hugo, typically suffering wind damage and removal. Such cladding

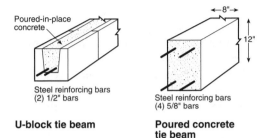

U-block tie beam | **Poured concrete tie beam**

(Poured-in-place concrete; Steel reinforcing bars (2) 1/2" bars — U-block tie beam)

(8"; 12"; Steel reinforcing bars (4) 5/8" bars — Poured concrete tie beam)

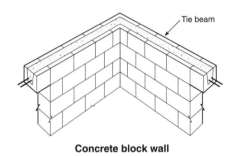

Tie beam

Concrete block wall

8.12. Reinforced tie beam (bond beam) for concrete block walls, to be used at each floor level and at roof level around the perimeter of the exterior walls.

should not be used on any building more than 30 feet tall along the coast of the Carolinas (appendix D, refs. 39, 161). Both brick veneer and wallboard cladding types of construction should be avoided along the coast. Lightweight preengineered metal walls also did not perform well in Hugo's winds, but these types of material generally are not used for residential buildings.

Windows and large glass areas should be protected, especially those that face the ocean. Many newer coastal houses have large areas of glazing. Windows and doors fail due to positive pressure and suction, and often are the weak link in the integrity of a structure. Objects blown through a window during a storm cause dangerous flying glass as well as weakening structural resistance. Wind-blown sand can very quickly frost a window and thereby decrease its aesthetic value. Both of these problems can be avoided if the house has storm shutters. Check to see that it does, and that they are functional.

Consult a good architect or structural engineer for advice if you are in doubt about any aspect of a house. A few dollars spent for wise counsel may save you from later financial grief.

To summarize, a beach house should have (1) roof tied to walls, walls tied to foundation, and foundation anchored to the earth (the connections are potentially the weakest link in the structural system); (2) a shape that resists storm forces; (3) shutters for all windows, but especially those facing the ocean; (4) floors high enough (sufficiently elevated) to be above most storm waters (usually the 100-year still-water flood level plus 3–8 feet to account for wave height); (5) piles that are of sufficient depth or posts embedded in concrete to anchor the structure and withstand erosion; and (6) piling that is well braced.

What Can Be Done to Improve an Existing House?

If you presently own a house or are contemplating buying one in a hurricane-prone area, you will want to know how to improve its ability to protect its occupants. Find the excellent government publication titled *Wind-Resistant Design Concepts for Residences,* by Delbart B. Ward (appendix D, ref. 175). Of particular interest are the sections on building a refuge shelter module within a residence (fig. 8.13). Also noteworthy are two supplements to this publication (appendix D, ref. 176), which deal with buildings larger than single-family residences and may be of interest to the general public, especially residents in urban areas. These works provide a means of checking whether the responsible authorities are doing their jobs to protect schools, office buildings, and apartments. Several other pertinent references are listed in appendix D as well.

Suppose your house is resting on blocks but not fastened to them, and thus is not adequately anchored to the ground. Can anything be done? One solution is to treat the house like a mobile home and screw ground anchors into the ground to a depth of 4 feet or more and then fasten them to the underside of the floor systems. Figures 8.14 and 8.15 illustrate how ground anchors can be used.

The number of ground anchors needed will be different for houses and mobile homes, because each is affected differently by the forces of wind and water. Note that recent practice is to put these commercial steel-rod anchors in at an angle in order to align them better with the direction of the pull. If a vertical anchor is used, the top 18 inches or so should be encased in a concrete cylinder about 12 inches in diameter. This prevents the top of the anchor rod from bending or slicing through the wet soil from the horizontal component of the pull.

Diagonal struts, either timber or pipe, may also be used to anchor a house that rests on blocks. This is done by fastening the upper ends of the struts to the floor system, and the lower ends to individual concrete footings substantially below the surface of the ground. These struts must be able to take both tension and compression, and should be tied into the concrete footing with anchoring devices such as straps or spikes.

If the house has a porch with exposed columns or posts, you should be able to install tie-down anchors on their tops and bottoms. Steel straps should suffice in most cases.

When accessible, roof rafters and trusses should be anchored to the wall system. Usually the roof trusses or braced rafters are sufficiently exposed to make it possible to strengthen joints (where two or more members meet) with collar beams or gussets, particularly at the peak of the roof (fig. 8.11).

A competent carpenter, architect, or structural engineer can review the house with you and help you decide what modifications are most practical and effective. Do not be misled

by someone who resists new ideas. One builder told a homeowner, "You don't want all those newfangled straps and anchoring devices. If you use them, the whole house will blow away, but if you build in the usual manner [with members lightly connected], you may lose only part of it." In fact, of course, the very purpose of the straps is to prevent any or all of the house from blowing away. The Standard Building Code (previously known as the Southern Standard Building Code and still frequently referred to by that name) says, "Lateral support securely anchored to all walls provides the best and only sound structural stability against horizontal thrusts, such as winds of exceptional velocity" (see appendix D, ref. 162). The cost of connecting all ele-

8.13. Bathroom shelter module. *Source:* Modified from *Wind-Resistant Design Concepts for Residences* (appendix D, ref. 175).

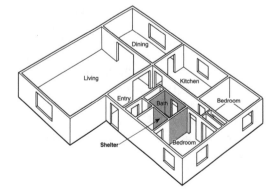

ments securely adds very little to the cost of the frame of the dwelling, usually under 10 percent, and a very much smaller percentage to the total cost of the house.

If the house has an overhanging eave and there are no openings on its underside, it may be feasible to cut openings and screen them. These openings keep the attic cooler (a plus in the summer) and may help to equalize the pressure inside and outside the house during a storm with a low-pressure center.

Another way a house can be improved is to modify one room so that it can be used as an emergency refuge in case you are trapped in a major storm. (This precaution is *not* an alternative to evacuation prior to a hurricane.) Examine the house and select the best room to stay in during a storm. A small windowless room such as a bathroom, utility room, den, or storage space is usually stronger than a room with windows. A sturdy inner room with more than one wall between it and the outside is safest. The fewer doors the better; an adjoining wall or baffle wall shielding the door adds to the protection.

Consider bracing or strengthening the interior walls. This may require removing the surface covering and installing plywood sheathing or strap bracing. Where wall studs are exposed, bracing straps offer a simple way to achieve needed reinforcement against the wind. These straps are commercially produced and are made of 16-gauge galvanized metal with prepunched holes for nailing. These

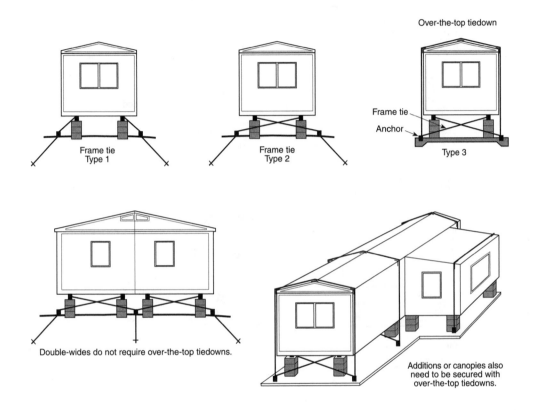

Frame tie
Type 1

Frame tie
Type 2

Over-the-top tiedown

Frame tie

Anchor

Type 3

Double-wides do not require over-the-top tiedowns.

Additions or canopies also
need to be secured with
over-the-top tiedowns.

8.14. Tie-downs for mobile homes. *Source:* Modified from *Protecting Mobile Homes from High Winds* (appendix D, ref. 198).

should be secured to studs and wall plates as nail holes permit (figs. 8.10, 8.11). Bear in mind that they are good only for tension.

If, after reading this, you agree that something should be done to your house, do it now. Do not put it off until the next hurricane or northeaster hits you!

Mobile Homes: Limiting Their Mobility

Because of their light weight and flat sides, mobile homes are exceptionally vulnerable to the high winds of hurricanes, tornadoes, and severe storms. High winds can overturn unanchored mobile homes or smash them into neighboring homes and property. Nearly 6 million Americans live in mobile homes today, and the number is growing. Mobile homes account for 20 to 30 percent of the single-family housing produced in the United States. High winds damage or destroy nearly 5,000 of these homes every year, and the number will surely rise unless protective measures are taken. As one man whose mobile home was overturned in Hurricane Frederic (1979) aptly put it, "People who live in flimsy houses shouldn't have hurricanes." Mobile homes suffered complete destruction during Hugo in areas where wind gusts exceeded 100 miles per hour.

Several lessons can be learned from past storms. First, mobile homes should be properly located. After Hurricane Camille (1969), observers noted that damage was minimized in mobile home parks surrounded by woods and where the units were close together; the damage was caused mainly by falling trees. In unprotected areas, however, many mobile homes were overturned and destroyed from the force of the wind. The protection afforded by trees is greater than the possible damage falling limbs may cause. Two or more rows of trees are better than a single row, and trees 30 feet or more tall give better protection than shorter ones. If possible, position the mobile home so that a narrow side faces the prevailing winds.

Locating a mobile home in a hilltop park

greatly increases its vulnerability to the wind. A lower site screened by trees is safer from the wind, but the elevation should be above storm-surge flood levels. A location that is too low, obviously, is subject to flooding. There are fewer safe locations for mobile homes than there are for stilt houses. *Manufactured Home Installation in Flood Hazard Areas* (appendix D, ref. 201) discusses flood hazards and manufactured homes.

A second lesson taught by past experience is that the mobile home must be tied down or anchored to the ground so that it will not overturn in high winds (figs. 8.14, 8.15; table 8.1). Simple prudence dictates the use of tie-downs, and in many communities ordinances require it. Many insurance companies, moreover, will not insure mobile homes unless they are adequately anchored with tie-downs.

A mobile home may be tied down with cable, rope, or built-in straps, or it may be rigidly attached to the ground by connecting it to a simple wood-post foundation system. An alert owner of a mobile home park can provide permanent concrete anchors or piers to which hold-down ties can be fastened. In general, an entire tie-down system costs a nominal amount.

A mobile home should be properly anchored with both ties to the frame and over-the-top straps; otherwise it may be damaged by sliding, overturning, or tossing. The most common cause of major damage is the tearing away of most or all of the roof. When this hap-

Commercial adapters or mounting brackets to prevent cable or strap tiedowns from cutting into the mobile home

If commercial adapter is not available, use wood blocks to distribute pressure of cable.

Cable: Galvanized steel, min. dia. 7/32". Galvanized aircraft, min. dia. 1/4" and (7x19)

At least two cable clamps with nuts placed on live side of cable.

Wire rope thimble

Closed eye

Drop-forged turnbuckle sized to equal breaking strength of rope

Top of anchor

Turnbuckles with hook ends should not be used. They can bend open under high wind loads.

8.15. Hardware for mobile home tie-downs. *Source:* Modified from *Protecting Mobile Homes from High Winds* (appendix D, ref. 198).

pens, the walls are no longer adequately supported at the top and are more likely to collapse. Total destruction of a mobile home is more likely if the roof blows off, especially if the roof blows off first and then the home overturns. The necessity for anchoring cannot be overemphasized: single mobile homes up to 14 feet wide require both over-the-top tie-downs to resist overturning and frame ties to

resist sliding off the piers. "Double-wides" do not require over-the-top ties, but they do require frame ties.

Mobile home owners should be sure to obtain a copy of the booklet *Protecting Mobile*

Homes from High Winds (appendix D, ref. 198), which treats the subject in more detail. The booklet lists specific steps to take on receiving a hurricane warning and suggests a type of community shelter for a mobile home park. It also includes a map of the United States that indicates the areas subject to the strongest sustained winds.

High-Rise Buildings: The Urban Shore

Any high-rise you see on the beach was probably designed by an architect and a structural engineer who were presumably well qualified and aware of the requirements for building on the shoreline. This does not mean that the requirements were followed! Tenants of such a building should not assume that it is invulnerable to storms. Many people living in two- and three-story apartment buildings were killed when the buildings were destroyed by Hurricane Camille in Mississippi in 1969. Storms have smashed five-story buildings in Delaware. Hugo did extensive damage to moderate- and high-rise buildings. The larger high-rises have yet to be thoroughly tested by a major hurricane.

The first aspect of high-rise construction that a prospective apartment dweller or condo owner must consider is the type of piling used. High-rises near the beach should be built so that the building will remain standing even if the foundation is severely undercut during a storm. It is well known in construction circles

Table 8.1   Tie-Down Anchorage Requirements

| Wind velocity (mph) | 10- and 12-Foot-Wide Mobile Homes | | | | 12- and 14-Foot-Wide Mobile Homes, 60–70 Feet Long | |
| | 30–50 Feet Long | | 50–60 Feet Long | | | |
	Number of frame ties	Number of over-the-top ties	Number of frame ties	Number of over-the-top ties	Number of frame ties	Number of over-the-top ties
70	3	2	4	2	4	2
80	4	3	5	3	5	3
90	5	4	6	4	7	4
100	6	5	7	5	8	6
110	7	6	9	6	10	7

Sources: Protecting Mobile Homes from High Winds (appendix D, ref. 198); and *Suggested Technical Requirements for Mobile Home Tie-Down Ordinances* (appendix D, ref. 199).

that less scrupulous builders sometimes take shortcuts and do not drive the piling deep enough. Just as important as driving the piling deep enough to resist scouring and to support the load it must carry is the need to fasten piles securely to the structure they support. The connections must resist horizontal loads from winds and waves during a storm and also resist uplift from the same sources. It is a joint responsibility of builders and building inspectors to make sure the job is done right. In 1975, Hurricane Eloise exposed the foundation of a high-rise under construction in Florida and revealed that some of the piling was not attached to the building. This happened in Panama City, Florida, but such problems probably exist everywhere that high-rises crowd the beach.

Despite the assurances that come with an engineered structure, life in a high-rise has definite drawbacks that prospective tenants should know about and take into consideration. They stem from high wind, high water, and poor foundations.

Pressure from the wind is greater near the shore than it is inland, and it increases with height. If you are living inland in a two-story house and plan to move to the eleventh floor of a high-rise on the shore, you should expect five times more wind pressure than you are accustomed to. It can be a great and possibly devastating surprise.

The high wind pressure can actually cause unpleasant motion of the building. It is worthwhile to check with current residents of the high-rise to find out if it has undesirable motion characteristics. Residents of some buildings claim that the swaying is enough to cause motion sickness. More seriously, high winds can break windows and damage other property, and of course they can hurt people. Tenants of severely damaged buildings have to relocate until repairs are made.

Those interested in researching the subject further—even the knowledgeable engineer or architect who is engaged to design a structure near the shore—should obtain a copy of *Structural Failures: Modes, Causes, Responsibilities* (appendix D, ref. 178). Of particular importance is the chapter titled "Failure of Structures Due to Extreme Winds," which analyzes wind damage to engineered high-rise buildings from the storms at Lubbock and Corpus Christi, Texas, in 1970.

Another occurrence that affects a multifamily high-rise building more seriously than a low-occupancy structure is a power failure or blackout. Power failures are more likely along the coast than inland because of the more severe weather conditions associated with coastal storms. A power failure can cause great distress. For example, people can be caught between floors in an elevator. Think of the mental and physical distress after several hours of confinement, compounded by the roaring winds of a hurricane whipping around the building. It is not difficult to imagine the many inconveniences that can be caused by a power failure in a multistory building.

Fire is particularly hazardous in high-rises. Even recently constructed buildings seem to have problems. The television pictures of a woman leaping from the window of a burning building in New Orleans rather than be incinerated in the blaze are a horrible reminder from recent history. There have been a number of hotel fires over the last few years. Fire department equipment reaches only so high, and many areas along the coast are too sparsely populated to afford high-reaching equipment anyway. Fire and smoke travel along ventilation ducts, elevator shafts, corridors, and similar passageways. The situation can be corrected and the building made safer, however, especially if it is new. Sprinkler systems should be operated by gravity water systems rather than by powered pumps (because of possible power failure); gravity systems use water from tanks higher up in the building. Battery-operated emergency lights that come on when the other lights fail, better fire walls and automatic sealing doors, pressurized stairwells, and emergency-operated elevators in pressurized shafts will all contribute to increasing safety. Unfortunately, these improvements cost money, and that is why they are often omitted unless required and enforced by the building code.

**Modular Unit Construction:
Prefabricating the Urban Shore**

The method of building a house, duplex, or larger condominium structure by fabricating modular units in a shop and assembling them at the site is becoming popular in shoreline developments. The largest prefabricated structures are commonly two to three stories tall and may contain a large number of living units.

Modular construction makes good economic sense, and there is nothing inherently wrong in this approach to coastal construction. The same methods have been used in the manufacturing of mobile homes for years, although final assembly of mobile homes is done in the shop rather than in the field. Doing as much of the work as possible in a shop can save considerable labor and cost. The workers are not affected by outside weather conditions; they can be paid by piecework, enhancing their productivity; and shop work lends itself to labor-saving equipment such as pneumatic nailing guns and overhead cranes.

If the manufacturer desires it, shop fabrication can permit higher quality. Inspection and control of the whole construction process are much easier. For instance, there is less hesitation about rejecting a poor piece of lumber when a good supply is nearby than when there is just so much lumber available on the building site.

On the other hand, because so much of the work is done out of the sight of the buyer, the manufacturer has the opportunity to take shortcuts if he is so inclined. Modules' wiring, plumbing, ventilation, and heating and air conditioning may be installed at the factory by unqualified personnel, and it is possible that the resulting inferior work is either not inspected at all or is inspected by an unconscientious or inept inspector. Therefore, it is important to consider the following points. Were wiring, plumbing, heating and air conditioning, and ventilation installed at the factory or at the building site? Were the installers licensed and certified? Was the work inspected both at the factory and on the construction site?

Most important, is the modular dwelling unit built to provide safety in the event of a fire? Just a few of the many safety features that should be present are two or more exits, stairs remote from each other, masonry fire walls between units, noncombustible wall sheeting, and compartmentalized units so that if fire does occur it will be confined to one unit. In general it is very desirable to check the reputation and integrity of the manufacturer just as you would when hiring a contractor to build your individual house on-site. The acquisition of a modular unit should be approached with the same caution used for other structures. If you are contemplating purchasing a modular dwelling units, you would be well advised to take the following steps:

1. Check the reputation and integrity of the developer and manufacturer.
2. Check to see if the developer has a state contractor's license.
3. Check the state law on who is required to approve and certify the building.

4. Check that building codes are enforced.
5. Check to see if the state fire marshal's office has indicated that the dwelling units comply with all applicable codes. Also check to see if this office makes periodic inspections.
6. Check to see that smoke alarms have been installed, that windows are the type that can be opened, that the bathroom has an exhaust fan, and that the kitchen has a vent through the roof.

As with all other types of structures, also consider site safety and escape route(s) for the location.

What Should Be Done?

The public's growing desire for less "big government" and less regulation has left coastal property more and more vulnerable. Vulnerable not just to natural hazards but also to weak or unenforced building codes, to construction cost reduction through the use of minimal materials that meet minimum code requirements at best, and to rushed construction, which can lead to poor-quality workmanship or, even worse, to shortcuts that render buildings unsafe. The adage "Buyer beware!" goes double for coastal property.

Coastal property owners and residents should insist that state and local governments follow the recommendations of postdisaster investigation teams. For example, most of the post-Andrew recommendations (appendix D, ref. 183) apply to the Carolinas, particularly the following:

1. The quality of workmanship needs to be improved. Both the construction industry and the building inspection enforcement arm need to better train their personnel.
2. Building codes must be improved and be better enforced.
3. Guidance on correct methods of transferring loads must be provided to building contractors.
4. Licensed design professionals should have more participation in the inspection of construction.
5. Inspector supervision should be improved.

And these activities should be ongoing and evaluated after each storm. Many of the same recommendations made after Hurricane Andrew had been made—and ignored—after Hurricane Hugo (appendix D, ref. 40). When will we learn?

A Final Consideration

Although less likely than a hurricane or northeaster, the threat of a damaging earthquake in South Carolina is very real. Fortunately, most of the construction recommendations to mitigate wind and flood damage also improve a building's ability to withstand earthquakes. Chapter 9 addresses earthquake-resistant construction specifically.

9 Earthquake-Resistant Design and Construction

Prudent construction practices in the South Carolina coastal zone focus on mitigating the damage from wind, wave, and flood forces. Earthquake-resistant design is either overlooked or ignored. Yet, each earthquake provides new insights on how design and construction techniques influence the level of damage to houses and other buildings. For example, the 1994 earthquake centered at Northridge, California, revealed that certain techniques for anchoring outside veneer to the framing of houses were inadequate, and that stucco applied over wall studs did not meet expectations for resisting seismic forces.

Each new earthquake reinforces two fundamental concepts about seismic-resistant design and construction: (1) ground shaking causes more than 90 percent of all earthquake damage to houses, and (2) earthquake forces will seek out and find any structural weakness in a house or building. These two principles were confirmed during the 1989 Loma Prieta earthquake. Houses built on soft soil with shallow groundwater (conditions not unlike those in the coastal Carolinas) as far as 60 miles from the earthquake's epicenter suffered significant damage. The primary damage in that quake was to houses with unreinforced masonry foundations and those not anchored to their foundations.

Few people will claim that they can build an earthquake-proof house. However, if certain fundamentals of design and construction are applied, the likelihood and extent of damages from earthquake forces will be significantly reduced. This chapter provides a basic understanding of how structures react to earthquake forces and points out design and construction concepts that should be considered or understood by anyone purchasing or constructing a house in an earthquake-prone area. Many of these techniques are applicable to and complement the design requirements for high wind and other hazards. In most cases, application of these techniques is not very costly, and it results in a better-built structure with a longer life.

How Buildings React to Earthquake Forces

Earthquakes create forces in structures because of inertia, the tendency for a body at rest (in this case the house or building) to resist motion, and for a body in motion to remain in motion. The forces that act on the structure depend on the direction of the ground motion caused by the earthquake; they may act side to side (horizontal) and/or up and down (vertically). During an earthquake, a structure resists the initial movement of the ground, but as the ground shaking continues, the structure eventually begins to move back and forth with the ground until the shaking stops and the earthquake is over. The larger earthquake-related forces are usually horizontal or lateral forces acting parallel to the ground. In houses, the effects of vertical forces are normally considered to be offset by the weight of the building. Figure 9.1 shows the inertial effects on a building frame caused by back-and-forth motion parallel to the ground. Similar effects would occur if the ground were stationary and a horizontal force was applied back and forth at the roof line.

Elements of Earthquake-Resistant Design and Construction

Siting

Structures in the South Carolina coastal zone are subject to several forms of risk connected with earthquakes. Besides the obvious damage that would be caused by fault rupture and/or ground shaking, buildings face the collateral problems of liquefaction and subsidence, both of which occurred in the 1886 Charleston earthquake. We have no record of how these processes affected the barrier islands, but it is critical that anyone building or buying a home on a barrier island understand the nature of the ground under and around the house. Site evaluation often requires professional guidance and consultation, but it is well worth the cost and effort. Obvious areas to avoid are marshes and other water-saturated soils, organic fills and alluvium, and loose, shifting soils. The sand soils of barrier islands with shallow water tables may liquefy in an earthquake.

The best sites to consider are on stable and solid rock formations, but these generally

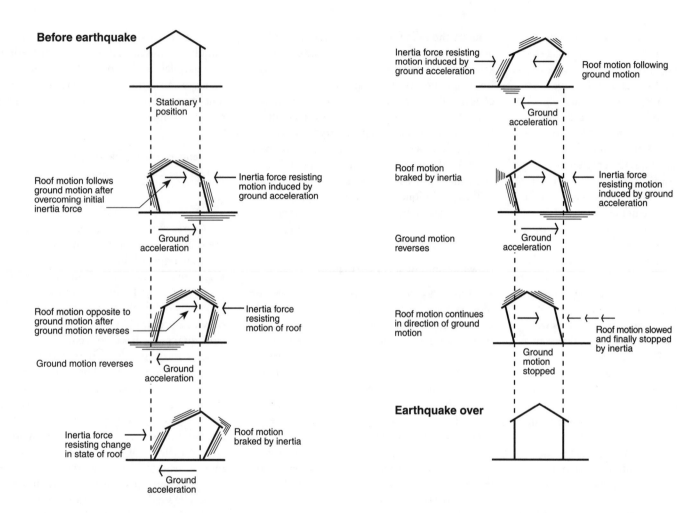

Before earthquake

Stationary position

Roof motion follows ground motion after overcoming initial inertia force

Inertia force resisting motion induced by ground acceleration

Ground acceleration

Roof motion opposite to ground motion after ground motion reverses

Inertia force resisting motion of roof

Ground motion reverses

Ground acceleration

Inertia force resisting change in state of roof

Roof motion braked by inertia

Ground acceleration

Inertia force resisting motion induced by ground acceleration

Roof motion following ground motion

Ground acceleration

Roof motion braked by inertia

Inertia force resisting motion induced by ground acceleration

Ground motion reverses

Ground acceleration

Roof motion continues in direction of ground motion

Roof motion slowed and finally stopped by inertia

Ground motion stopped

Earthquake over

9.1. Inertial effects on the frame of a building caused by a back-and-forth motion parallel to the ground.

aren't available along the South Carolina coast. Inland, deep bedrock provides the least hazardous conditions and minimizes earthquake damage to dwellings. Natural sites that consist of firm, consolidated deposits of well-drained soil, either flat or sloping, and stable hillside slopes are acceptable. Less desirable sites may need to be improved by removing problem materials and replacing them with engineered fill, by stripping and terracing sloped sites, and/or by providing additional drainage.

Design and Construction

Dr. Karl Steinbrugge, the father of modern earthquake engineering, coined the phrase that has become the standard opening line for many an earthquake speech: "Earthquakes don't kill people, buildings do." Modern building codes and improved construction techniques and materials have spared America the enormous loss of life from earthquakes such as those as experienced several years ago in Armenia and more recently in the 1995 Kobe, Japan, earthquake, in which almost 5,000 people died in collapsed structures. Nevertheless, we should not become complacent. Attention to earthquake-resistant design and use of good construction techniques can reduce injury and property loss, as well as loss of life.

In general, one- and two-family dwellings, because of their design characteristics and ma-terials, tend not to collapse in earthquakes. This applies particularly in the western United States, where wood-frame construction is the predominant housing type. Wood-frame construction is known to perform better in earthquakes than other construction such as unreinforced masonry, although this does not mean that it escapes significant damage. And this damage often makes the structure uninhabitable, causing both economic and social trauma to the occupants.

Most conventional homes can resist the horizontal forces of earthquakes (or wind) because they are built in a boxlike configuration. The plainer the design, the more stable the structure. The box configuration provides resistance by means of roof and floors (horizontal) and walls (vertical). Roofs and floors are considered to be horizontal diaphragms. A horizontal diaphragm can be compared to a steel beam that has a top and bottom flange and web, with the web oriented in a horizontal plane. In house construction, the exterior wall's top plates act as flanges and the roof sheathing functions as the web (fig. 9.2). For the roof to act most effectively as a horizontal diaphragm, it should be of a simple, regular shape with unbroken planes. A flat, pitched, or gabled roof is the best configuration.

Floor diaphragms are also most effective when they are simple in shape and designed in a geometric pattern. Shape and design are particularly important in a first-floor diaphragm spanning a basement opening and on cripple-wall foundations. Cripple walls are usually wooden-stud walls built on top of an exterior foundation to support a house and create the crawl space. It is best to apply a conventional design combined with a corresponding symmetrical pattern for the shear walls or shear panels (walls that resist horizontal forces; see below).

Floors and roofs often have to be penetrated for practical purposes such as duct shafts, and for aesthetic purposes such as skylights. The size and location of such openings must be carefully considered because they can have a critical impact on how effectively the diaphragm will function during an earthquake. Floors and roofs must be securely anchored to walls at the perimeter and at intermediate locations.

Walls that resist horizontal forces are known as shear walls or shear panels, the latter being resistant elements that might be part of a longer wall. Shear walls can be viewed as upright beams on a fixed base comparable to a vertical cantilever beam (a beam supported at one end only), with the end studs of the sheathed portion acting as flanges and the sheathing between end studs acting as the web. During an earthquake, ground motions enter the structure and create inertial forces which move the floor diaphragms. This movement is resisted by the shear walls, and the forces are transmitted back down to the foundation.

The shape and size of the shear walls are important design criteria. Walls that are too tall

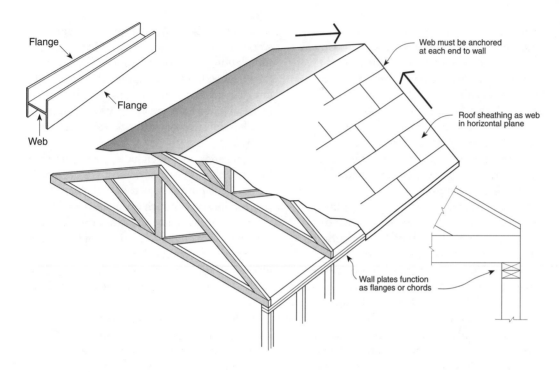

Flange

Flange

Web

Web must be anchored
at each end to wall

Roof sheathing as web
in horizontal plane

Wall plates function
as flanges or chords

9.2. Roofs and floors must act as horizontal diaphragms, much like a steel beam does.

or narrow tend to tip over before they slide. Because earthquakes can create forces in any direction, hold-downs must be placed at each end of a shear wall or panel. Shear wall patterns are configured so that opposite and parallel pairs of walls resist loads in a single direction, and the exterior walls help the house resist twisting or racking (figs. 9.3, 9.4). Shear

walls must be located at or near the boundaries of the roof and floor diaphragms to be effective; that is, at each exterior wall.

In order to ensure that construction is earthquake resistant, it is essential to provide a continuous load path from roof to foundation in order to dissipate earthquake forces. Good connections between the resisting elements provide the continuity to ensure an uninterrupted path and tie the building components together (figs. 9.5, 9.6).

One of the most common types of earth-

quake damage to homes occurs when an improperly anchored structure shifts off its foundation. This can rupture gas lines, resulting in fire, and cause exterior damage to walls and windows as well as interior content damage. (It is also very expensive to lift a house and put it back on its foundation.)

In order to keep the house from sliding off its foundation, all the structural components (roof, shear walls, and floors) must be securely tied to the foundation. This structural integrity is most effectively achieved by anchoring the sill plate to the foundation. (The sill plate is the wooden board that sits directly on top of the foundation and secures the house to the foundation.) The anchor bolts must be located accurately on the center line of the mud sills, and the mud sill plates must be secured to and consistently spaced along the foundation. To check whether a house is properly anchored, look in the crawl space for the heads of the anchor bolts. You should be able to see them installed every 4 to 6 feet along the sill plate. Other forms of connectors, such as steel plates, may have been used to connect the frame to the foundation.

Houses on foundations of unreinforced masonry such as brick or concrete block also tend to move off their foundations during earthquake shaking. Brick foundations should have visible anchor bolts or steel reinforcing bars between the inner and outer faces. Concrete block foundations should be reinforced with anchored steel bars embedded in the grout fill

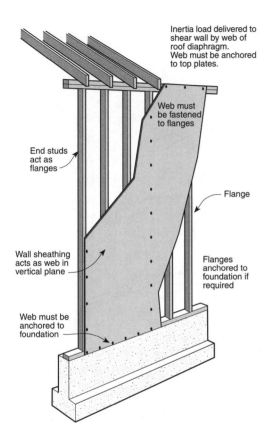

Inertia load delivered to shear wall by web of roof diaphragm. Web must be anchored to top plates.

Web must be fastened to flanges

End studs act as flanges

Flange

Wall sheathing acts as web in vertical plane

Flanges anchored to foundation if required

Web must be anchored to foundation

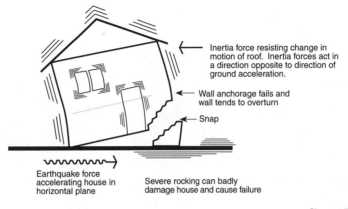

Inertia force resisting change in motion of roof. Inertia forces act in a direction opposite to direction of ground acceleration.

Wall anchorage fails and wall tends to overturn

Snap

Earthquake force accelerating house in horizontal plane

Severe rocking can badly damage house and cause failure

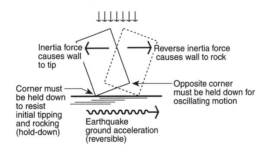

Inertia force causes wall to tip

Reverse inertia force causes wall to rock

Corner must be held down to resist initial tipping and rocking (hold-down)

Opposite corner must be held down for oscillating motion

Earthquake ground acceleration (reversible)

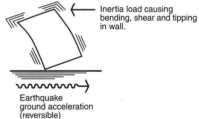

Shear walls in rocking mode can continue to function as a beam carrying inertia loads of roof. However, rocking can damage wall or cause wall to shift off foundation.

Inertia load causing bending, shear and tipping in wall.

Earthquake ground acceleration (reversible)

9.3. Above: Requirements for an earthquake-resistant wall.

9.4. Right: A secondary effect of earthquake forces is overturning of walls. This is best resisted by "hold-downs," anchors to the foundation.

of the blocks. If the cells of the concrete blocks are hollow, the foundation is probably unreinforced and could fail, allowing the house to shift during an earthquake.

Building design or construction that incorporates cripple walls does not perform well in earthquake-prone areas if the walls are not

properly braced. Plywood panels or diagonal wood sheathing forms the best braces for cripple walls.

Houses built on concrete slabs are generally considered to include a degree of seismic resistance because they do not have cripple walls and are bolted to the foundation. In most cases, such houses have foundation anchor bolts installed during construction, and these anchor bolts are usually visible. If they are not, you can usually tell if they are there by examining the walls connecting the garage to the house.

Masonry chimneys are extremely vulnerable to earthquake forces. These heavy and brittle structures are usually constructed to be free standing above the roof line and pose special problems when used with flexible buildings such as wood-frame homes. The chimney should be vertically supported on a reinforced concrete pad, and the chimney walls should have a minimum of horizontal and vertical reinforcing. Separation of the chimney from the structure during ground shaking is the predominant damage pattern. To prevent this, the chimney should be anchored to the house framing with ties at every diaphragm level (floor, ceiling, and roof) that are embedded in the masonry and strapped around the vertical reinforcing. Depending on the size and height of the chimney relative to the frame, additional ties may be needed. Also, metal flues should be considered as replacement for masonry. Determining whether a chimney is susceptible to earthquakes is not easy. Tall, slender chimneys that dramatically exceed the plane of the roof tend to be the most vulnerable. Inspection by a professional often is necessary.

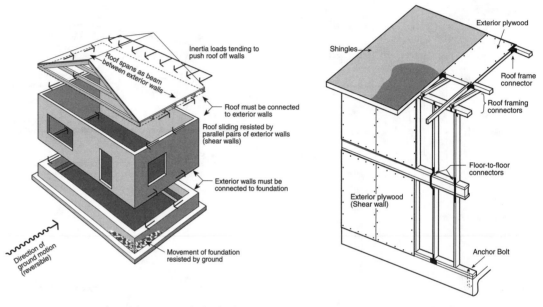

9.5. Continuity is the guiding principle for both wind-resistant and earthquake-resistant construction.

9.6. The recommended wood-frame construction has a continuous load path.

Nonstructural Components

The inside of your home—its contents—may be a greater threat to your safety during an earthquake than collapsing walls and roof. Few people bother to consider fixtures, appliances, and other objects that can fall over and cause personal injury or a fire hazard. Several excellent documents are available that can be used to assess interior hazards around your house. Two in particular are *Checklist of Nonstructural Earthquake Hazards* (appendix D, ref. 208) and *Reducing the Risks of Nonstructural Earthquake Damage: A Practical Guide* (appendix D, ref. 204). Both documents are available free of charge.

Two particularly important nonstructural hazards are improperly braced water heaters and free-standing stoves. Water heaters not securely fastened to the wall may topple over during an earthquake and cause extensive water damage, and fires can be ignited when the connecting gas or electric lines break. Ensuring that this doesn't happen is relatively simple and inexpensive. Metal straps or braces should encircle the heater and be secured by bolts to the wall. These bolts should go into studs or concrete, but not into drywall because it can't support the weight of the heater during intense ground shaking. Also, flexible pipes connecting the gas and water lines perform better during ground shaking than more rigid aluminum or copper pipes.

Free-standing stoves are common sources of residential heat. In most cases, fire codes set the requirements for the construction and siting of these units. One of the requirements is that there be significant clearance around the stove, which often means that the stove is unsupported on all four sides, making it extremely vulnerable to sliding or overturning during an earthquake. To prevent a fire hazard, the stove should be anchored to the floor and the stovepipe sections secured to prevent separation. It is important that the materials used to do this anchoring and bracing be heat resistant.

Manufactured Homes

Manufactured homes are seldom destroyed by earthquakes, but even a moderate earthquake can separate a mobile home from its foundation or piers and cause significant damage to awnings, decks, steps, skirting, and other ac-

cessory structures attached to it. Mobile homes often rotate on their foundations in an earthquake so that the doorway no longer opens onto the deck, porch, or steps. More than one resident has survived a quake only to step out of the door into unexpected space and take a short fall to injury. Fires often result when gas line connections to the structure or to the appliances inside it rupture. Reinstallation of damaged mobile homes can be quite expensive.

Bracing systems are available to secure the manufactured home so that it will not fall to the ground. To prevent earthquake-related fires, install flexible gas lines and secure all gas-burning appliances in place with strapping to ensure that they won't be dislocated by the ground shaking.

Codes and Current Practices

The foregoing concepts represent the very basics of earthquake-resistant design for residential buildings. Residential design and construction is one of the most rapidly changing and innovative industries in the nation today. The frequent changes in construction technology for the housing market make it especially important for buyers and residents alike to carefully review local building codes and consult the reference materials included in appendix D. Incorporating earthquake resistance into the design and construction of a residence most often requires specialized advice, and we

encourage you to utilize the services of design and engineering professionals. Contact FEMA for the most recent earthquake-resistant construction guidelines. In addition, recent actions on the part of the model code groups indicate that they are considering adopting the Council of American Building Officials (CABO) One and Two Family Dwelling Code as the nationally accepted code for such construction. The seismic provisions of the CABO code should be substituted for the current provisions included in the Standard Building Code, the model building code most often applied in South Carolina.

Awareness of seismic hazards is a matter of vital interest to individuals as well as communities, and individuals should take responsibility for reducing the risks associated with such hazards. Mitigation is accomplished during the siting, design, and construction of buildings, and the safety of our hospitals, schools, and theaters is a function of public awareness. Public expressions of interest and concern make everyone more sensitive to the seismic hazard and more interested in taking the proper actions to reduce the associated risks (see appendixes A and D).

If you own or plan to buy coastal property in South Carolina, you should be aware of your responsibilities under the current law with respect to development and land use. This chapter includes a partial list of land-use regulations applicable to the South Carolina coast and a discussion of the evolution of coastal management from the federal level to state and local levels. The explanations provided here are general; appendix B lists the agencies that will supply more specific information. In most cases you should contact the South Carolina Department of Health and Environmental Control (DHEC), Office of Ocean and Coastal Resource Management (OCRM) (formerly the South Carolina Coastal Council) or your county or municipal planning office if you want to know more.

The National Flood Insurance Program

One of the most significant legal pressures applied to encourage land-use planning and management in the coastal zone is the National Flood Insurance Program (NFIP) administered by the Federal Insurance Administration (FIA), a part of the Federal Emergency Management Agency (FEMA). The National Flood Insurance Act of 1968 (P.L. 90-448) as amended by the Flood Disaster Protection Act of 1973 (P.L. 92-234) created the NFIP in order to (1) identify and map all flood-prone areas, (2) provide affordable flood insurance to the public through a federal–private industry program, and (3) en-

courage land-use planning in flood-prone areas in order to minimize the need for disaster relief.

The NFIP requires participating communities to meet certain conditions to make homeowners eligible to purchase flood insurance. Owners of property in flood zones who do not purchase such insurance will not, in the event of a flood, receive most forms of federal financial assistance. For example, home mortgage insurance from the FHA and VA and aid from the Small Business Administration or Department of Agriculture are available only when the individual and community involved have complied with the requirements of the law. Federal funds for shoreline engineering, waste disposal, or water treatment systems in flood zones are not likely to be approved for nonparticipating communities. Communities must adopt certain land-use measures to make flood insurance available at reasonable rates. Although the law was conceived for and is often associated with river floodplains, it also applies to barrier islands and coastal areas subject to storm-surge flooding and inundation by waves.

The initiative for qualifying for the program rests with the community, which must contact FEMA (see appendixes B and D). Once the community adopts initial land-use measures and applies for eligibility, FEMA designates the community as eligible for subsidized insurance under the emergency program. When a community qualifies for the emergency program, a

Flood Hazard Boundary Map (FHBM) that approximately delineates flood-prone areas goes into effect. Ultimately, FEMA conducts an engineering study and provides the community with a Flood Insurance Rate Map (FIRM) which offers more detailed delineations of flood hazard areas and data on 100-year-flood elevations (fig. 10.1). (A 100-year flood is a flood having a 1 percent annual probability of occurring.) Flood Insurance Rate Maps also designate flood insurance zones. These zones include *A-zones,* areas that would be flooded by a 100-year flood, and *V-zones* (coastal high-hazard areas), locations that would be penetrated by waves on top of the storm-surge flood. Before the FIRM is issued, flood insurance is available to existing and new structures at a flat subsidized rate. After issuance of the FIRM, all new construction and significantly improved older structures must meet elevation requirements to comply with local regulations and pay additional (true risk) rates for insurance coverage. As the costliness of the coastal flood experience becomes clearer, these rates are likely to rise.

Since October 1981, all residential buildings in V-zones must meet elevation requirements that accommodate potential wave height (usually 2–6 feet), and residents of these areas may be charged higher actuarial rates. Nonresidential structures not located in V-zones may not have to be elevated, but they must be floodproofed to be eligible for insurance. General eligibility requirements vary among el-

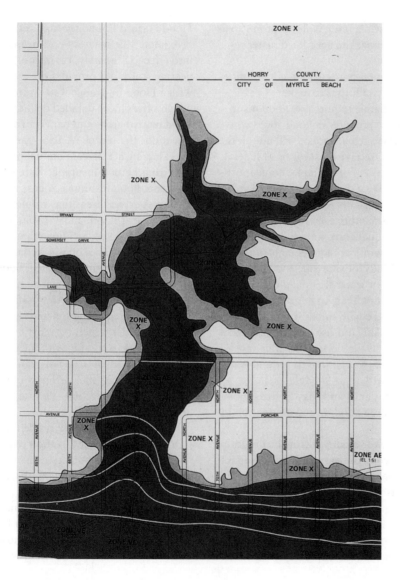

ZONE X

HORRY   COUNTY
CITY   OF   MYRTLE   BEACH

ZONE X

ZONE X

BRYANT   STREET

SOMERSET   DRIVE

ZONE AE

LANE

ZONE X   ZONE X

ZONE X

NORTH

ZONE AE

ZONE X   ZONE X

NORTH   NORTH   NORTH   NORTH

AVENUE   PORCHER

ZONE X   ZONE X   ZONE X

AVENUE   AVENUE

ZONE AE   ZONE AE   AVENUE   AVENUE   AVENUE   AVENUE   AVENUE   ZONE AE
(EL 15)

65TH   65TH   70TH   70TH   ZONE X

AE   ZONE VE

ZONE VE   ZONE VE

ZONE VE

10.1. A Flood Insurance Rate Map showing Myrtle Beach.

evated houses, mobile homes, and condominiums.

Although detailed flood risk analysis has been conducted to establish the degree of insurance risk in areas that would be flooded by a 100-year flood, a similar level of detail for V-zones was not available until the National Weather Service developed a computer simulation model called SLOSH (for Sea, Lake, and Overland Surges from Hurricanes). The SLOSH model is used to determine the still-water superelevation of storm waters caused by barometric pressure, hurricane intensity, wind speed, forward speed of the storm, and so forth; that is, the height and penetration of waves above the still-water surge elevations for the coast, taking into account hurricanes approaching from various directions (fig. 10.2). The model calculations require detailed bathymetric data for the nearshore area. These types of studies have been essential for developing hurricane evacuation plans in exposed coastal areas and have contributed greatly to reducing loss of lives as well as improving hurricane tracking and prediction capabilities.

As of 1993, the NFIP regulations addressed only flooding; however, rates of shoreline erosion/retreat are being considered for incorporation into the coastal hazard zone designations. Well-established historic shoreline erosion rates will be used to zone the

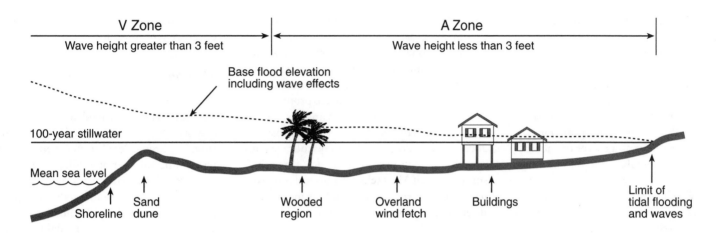

V Zone — Wave height greater than 3 feet

A Zone — Wave height less than 3 feet

Base flood elevation including wave effects

100-year stillwater

Mean sea level

Shoreline | Sand dune | Wooded region | Overland wind fetch | Buildings | Limit of tidal flooding and waves

10.2. The V-zone and A-zone in a coastal flood zone, as defined by FEMA. *Source:* Modified from *Managing Coastal Erosion* (appendix D, ref. 152).

nearshore. Coastal property will be classified as being in, for example, a 10-year, 30-year, or 60-year zone, depending on where the shoreline will be in 10, 30, or 60 years as predicted by projecting historic rates of erosion into the future. Based on such zoning, certain land-use practices will be allowed only in certain areas. Developers and individual property owners of shoreline property can expect that erosion rates will be part of the basis for future land-use regulation.

Property owners may purchase flood insurance from a licensed property agent, a state-approved broker, or an agent with a "write your own" (WYO) company specializing in flood insurance. The policy term is one year (three years under WYO). A single-family dwelling may carry a maximum of $185,000 on the building and $60,000 on the contents. The average cost of coastal flood insurance is $262, and $469 in the high-hazard zone.

The NFIP's revenues come from receipts from program operations, policy premiums, and U.S. Treasury borrowings and congressional appropriations. The money is kept in the Flood Insurance Fund.

Before building or buying a home, an individual should ask certain basic questions about flood insurance:

1. Is the community I'm locating in covered by the program?
2. If not, why not?
3. Is my building site above the 100-year flood level, plus wave height of 2 to 6 feet if it is located in a V-zone?
4. What are the structural requirements for my building?
5. What are the limits of coverage?

Most lending institutions and building inspectors are aware of mapped flood-prone areas, but it would be wise to confirm such information with the appropriate insurance representative or program office (see appendix B, "Insurance").

Government literature states that the NFIP would not have been necessary "had adequate and assured flood-insurance been available through the private insurance market." This is not meant to impugn the private insurance industry, however. The reason potentially interested companies are not in the flood insurance business is that they are too small, too regional, and unable to spread the risk over enough policyholders to make coverage feasible. Moreover, the engineering and infrastructure necessary to put the program in place, along with the need for floodplain management to keep rates affordable, are well beyond the capabilities of any single company.

Actually, the NFIP is not designed solely to "compensate" property owners in high-risk areas. It has a further, twofold, purpose. First, federal insurance serves as a carrot to encourage communities to regulate floodplain areas. Prior to the NFIP, only a small percentage of the nation's floodplains were regulated in any way. Second, it was intended to replace a system of disaster assistance that had proved ineffective. Under the former system the first few thousand loans were routinely forgiven; no strings were attached to loans in order to encourage sound development; and, of course, there were always casualty deductions to cushion losses. A federal insurance system was thought to be far more likely to reduce federal expenditures in the long run. This has not proven to be the case, because disaster relief funds have always been made available in some form.

Thus the NFIP was always intended to perform on a sound actuarial basis. Many poor decisions, based on lack of data, were made early in the program. Rates across the board were set too low. Only experience gained from more than 20 years of losses allowed the current rates to be adjusted to reflect risk more accurately.

As presently instituted, the NFIP still has flaws. Probably the most frequent complaint is that the boundaries of the FIRMs are inaccurate or difficult to use. There is an appeals system designed to solve this problem. Another claim is that flood insurance encourages individuals to build or buy homes in disaster-prone areas. Studies of insurance purchases are mixed on this score, but the availability of flood insurance certainly does nothing to discourage construction in high-risk areas. In fact, along the south shore of Massachusetts several homes have been destroyed and rebuilt as many as three times—all with the aid of federal flood insurance.

Perhaps the greatest obstacle to the success of the program is the uninformed individuals who stand to gain the most from it. One study that examined the public's understanding of flood insurance found that many people had little awareness of the flood threat or the cost of flood insurance. They viewed insurance as an investment with the expectation of a return rather than as a means of sharing the cost of natural disasters.

Prospective homeowners should be aware of moves in Congress to alter the structure of the flood insurance program. Already a much greater share of the program has been shifted away from the taxpayer to the property owner, causing a dramatic rise in premiums. Also, the Coastal Barrier Resources Act enacted by Congress in October 1982 made new structures built in certain areas after October 1983 ineligible for coverage (see the discussion of this law below). These trends in flood insurance coverage will continue.

The Upton-Jones Amendment

The Upton-Jones Amendment (section 1306[c]), which provided an incentive to relocate structures out of areas threatened by erosion, has been deleted from the NFIP. Even without it, however, it is still possible to get funds to relocate structures. Specific applications may be made to the NFIP. The applications are evaluated and assigned priorities, and when sufficient funds are available, the program finances relocation of the structure(s).

Some Flood Insurance Facts

Know the difference between a homeowner's policy and a flood insurance policy:

1. Flood insurance offers the potential flood victim a less expensive and broader form of protection than would become available through a postdisaster loan.
2. Flood insurance is a separate policy from homeowner's insurance. From the stand point of water damage, the latter covers only structural damage from wind-driven rain.
3. Flood insurance covers losses that result from the general and temporary flooding of normally dry land, the overflow of inland or tidal water, and the unusual and

rapid accumulation of runoff of surface water from any source.

Check to see if your property location has been identified as flood prone on the federal FIRM. If you are in a flood-prone area, you must purchase flood insurance in order to be eligible for any form of federal or federally funded financial, building, or acquisition assistance—that is, VA and FHA mortgages, Small Business Administration loans, and similar assistance programs. To locate your property on the FIRM, see your insurance agent. Also keep in mind that (1) you need a separate policy for each structure; (2) if you own the building, you can insure structure and contents, contents only, or structure only; and (3) if you rent, you need only insure the contents. A separate policy is required to insure the property of each tenant.

For flood insurance purposes, a condominium unit that is a traditional townhouse or row house is considered a single-family dwelling, and the individual units may be separately insured.

Mobile homes are eligible for coverage if they are on foundations, permanent or not, and regardless of whether the wheels were removed either at the time of purchase or while on the foundation.

Structures and other items not eligible for flood insurance include travel trailers and campers, fences, retaining walls, seawalls, septic tanks, outdoor swimming pools, gas and liquid storage tanks, wharves, piers, bulkheads, growing crops, shrubbery, land, livestock, roads, and motor vehicles.

One insurance broker cannot charge you more than another for the same flood insurance policy, because the rates are subsidized and set by the federal government. There is a five-day waiting period from the date of application until the coverage becomes effective.

The Coastal Barrier Resources Act

On October 18, 1982, the United States Congress passed the Coastal Barrier Resources Act (P.L. 97-348) designating what has been called the COBRA system. Its purpose is to minimize the wasteful spending of federal tax dollars for development-related activities in higher-risk areas on certain barrier islands located on the Atlantic and Gulf coasts, and to minimize loss of valuable fish and wildlife habitat resulting from undesirable development of these areas.

Specifically, COBRA designates areas where the federal government will not subsidize the costs of bridges, roads, and infrastructures such as sewer and water lines. This does not affect the federal government's involvement in current activities such as maintenance dredging, federal disaster aid, and Coast Guard activities.

You also should be aware that any structures built in the designated areas after October 1, 1983, are not eligible for federal flood insurance. The same restriction applies to substantial improvements on existing structures. The unavailability of flood insurance makes ownership in these areas a risky proposition, since even the best-built structures can be destroyed by large storms. Mortgage loans also are likely to be difficult to obtain.

Numerous barrier islands in South Carolina are affected by the provisions of this law, including Waites Island, Litchfield Spit, Pawleys Island (extreme ends), Debidue Beach (near Pawleys Inlet and North Inlet), Dewees Island (in part), Morris Island, Bird Key, Kiawah Island (near Stono Inlet and Captain Sams Inlet), Seabrook Island (near Captain Sams Inlet), Botany Island (in part), Botany Bay Island, Edingsville Beach, Otter Island, Harbor Island (in part), Pritchards Island, Little Capers Island, St. Phillips Island, Bay Point Island, and Daufuskie Island (in part). For more detailed information on the exact areas affected, check with the OCRM (see appendix B).

The Coastal Barrier Improvement Act

Passed on November 16, 1990, the Coastal Barrier Improvement Act (COBIA; P.L. 101-591) amended the Coastal Barrier Resources Act by expanding the COBRA system and broadening the definition of the term *coastal barrier* to include barriers composed of consolidated sediments. In addition to adding more protected acres along the South Carolina coast, COBIA added areas along the Atlantic and Gulf coasts, Puerto Rico, and the Virgin

Islands to the COBRA system. The number of units in the COBRA system was increased from 13 to 16, the total acreage covered went from 26,887 acres to 97,788 acres, and the total fastland acreage (land above mean high tide) covered increased from 4,379 acres to 7,705 acres.

Flood insurance is not available in COBIA units, which are designated on special maps published by FEMA and the U.S. Geological Survey that became effective August 3, 1992 (see appendix B).

The South Carolina Coastal Zone Program

The federal Coastal Zone Management Act of 1972 (P.L. 92-83) set in motion an effort by most coastal states to manage their shorelines and thereby conserve a vital national resource. A key requirement of the act is coastal land-use planning based on land classification and on identification and protection of critical areas.

In 1977 the South Carolina Coastal Zone Management Act (SCCZMA; Act 123) was passed in compliance with the federal act, qualifying South Carolina for federal aid in implementing effective coastal management (appendix D, ref. 134). The SCCZMA established the basis for the South Carolina coastal program that was approved by the state government in 1979. The South Carolina Coastal Council, now the Office of Ocean and Coastal Resource Management, was established to ad-

minister a comprehensive coastal management program to encourage protection and sound development of coastal resources. At the heart of the program is the permitting process, which requires the OCRM to approve certain activities in the four "critical areas" of the coastal zone (tidelands, coastal waters, beaches, and primary oceanfront sand dunes).

Any alteration of these critical areas through dredge and fill, drainage, removal of material, or construction of any of a wide variety of structures (e.g., buildings, docks, shore protective devices, and water systems) requires a permit from the OCRM. Failure to comply or violation of a permit can result in fines of up to $5,000, six months' imprisonment, or both for the first offense.

Anyone contemplating any kind of development in the coastal zone should contact the Charleston office of the OCRM (see appendix B) and request a copy of *Permitting Rules and Regulations* (appendix D, ref. 141) and an application form with instructions. The staff will provide assistance and guidance in completing applications and meeting the necessary requirements.

In the case of dredging and filling or construction along navigable waters, a permit is also required from the U.S. Army Corps of Engineers. For septic or water systems, a specific permit from the DHEC is required, but the OCRM has final authority over all permits if the project is in the critical zone. Fortunately, these dual permit procedures are coordinated,

and the OCRM staff can advise you if other permits are needed.

South Carolina's Beachfront Management Act

By the mid-1980s the proliferation of seawall applications at the Coastal Council had grown to alarming proportions. As stipulated in the rules and regulations for permits in critical areas, a seawall permit was granted whenever a habitable structure was threatened with erosion. A seawall permit was never a question of "if," only a matter of "when." Thus, in spite of the state's outstanding coastal management program, many parts of the South Carolina coast were being densely developed behind armored shorelines.

Near the end of 1987 the Blue Ribbon Committee of Beachfront Management was convened by former state senator James Waddell and assigned the task of evaluating the condition of South Carolina's beaches and the existing program for managing critical beach areas; in addition, the committee was asked to make recommendations. The committee's report (appendix D, ref. 156) was submitted to the legislature, and after some modification became the Beachfront Management Act on July 1, 1988. The Coastal Council was given two years to verify erosion rates and determine the exact locations of jurisdiction lines. The final Beachfront Management Act became effective on July 2, 1990. In between these two dates, in 1989, Hurricane Hugo paid a visit. Hugo tested the new law, and the law emerged unchanged.

The South Carolina Beachfront Management Act has drawn national attention and acclaim for its forward-looking management and its means of dealing with eroding shorelines. Under the act, no new seawalls may be built in South Carolina; existing walls may be maintained until they are destroyed. Other fundamental hallmarks of this law include regulations for setback areas along the beaches. The size of the setback is a function of the local erosion rate (40 times the annual rate). Building in setback areas is not prohibited, but it is regulated by the OCRM.

The State's Jurisdiction along Beaches

In the years following the adoption of the Coastal Zone Management Act it became evident that the state's jurisdiction was inadequate to protect beaches from overdevelopment. If dunes were present in a natural state, their landward trough defined the state's jurisdiction, the "critical line." But often the dunes had been removed by or for development or by natural forces. In these cases the critical line was to be drawn along a shoreline hardening structure or the "highest uprush of the waves." This was somewhat ambiguous. Worse yet, it provided no buffer zone or room to undo the mistakes of the past—where a migrating shoreline had been prevented from moving by a seawall. In other words, high-density develop-

ment was increasing in the worst possible places—where seawalls defined the end of the state's jurisdiction. Condominiums were being built just behind seawalls in parts of the Grand Strand and other areas.

The Beachfront Management Act's solution to the problem was twofold. First, the proliferation of seawalls was stopped; existing walls can remain but will be removed when they are destroyed. Second, a setback area was created along the South Carolina shores which accommodates shoreline retreat because its size is a function of the erosion rate. The setback zone provides room for a frontal dune to form or be built. Thus, the central flaw in South Carolina's ability to effectively manage its migrating shoreline was resolved by this excellent piece of legislation.

The most fundamental aspect of the Beachfront Management Act is its regulation of setback areas along South Carolina's coast. Setbacks are defined by baselines and setback lines, which are determined by the state. For a detailed explanation of setbacks, consult appendix C. Anyone desiring specific information should contact the OCRM (see appendix B for the address) or seek legal counsel.

Before baselines and setback lines are determined, an island's shorefront is divided into three specific erosion zones: standard, unstabilized inlet, and stabilized inlet. On the risk maps in chapter 6, these are zones denoted by *S, Iu,* and *Is,* respectively.

A standard (erosion) zone is a segment of

shoreline unaffected by inlets, tidal currents, or shoals. Erosion is relatively predictable and perpendicular to the shore.

An unstabilized inlet zone is a shoreline adjacent to an inlet where erosion is caused by or related to inlet processes. Unstabilized, or natural, inlets, though less predictable than standard zones, are predictable within limits.

A stabilized inlet zone is a segment of shoreline adjacent to an inlet stabilized by jetties, terminal groins, or other massive shoreline structures. Though a stabilized inlet is not supposed to migrate and its processes are inhibited, the adjacent shorelines do erode, accrete, overwash, and so on.

The type of erosion zone determines the location of the baseline. Once the baseline has been established, the state's official erosion rate is used to determine how far landward the setback line will be.

Baselines and Setback Lines

Along most shorelines, the baseline is the primary dune crest line. If no dune exists, then the location of the "ideal dune" crest—that is, where the dune would be under natural conditions—is calculated. Armored shorelines typically are deficient in beach sand. The methodology used to determine the "ideal dune" shifts the baseline landward to compensate for this deficit (fig. 10.3). On armored shorelines the baseline (where the dune "should" be) often runs behind habitable structures that sit in

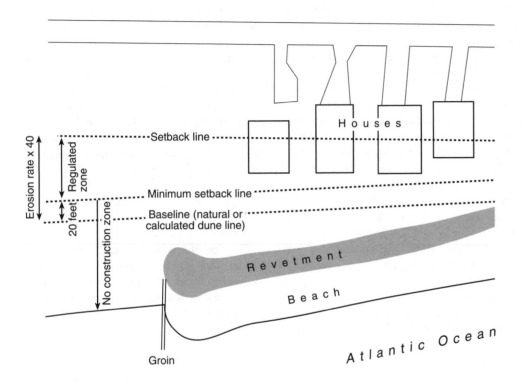

Erosion rate x 40

Regulated zone

20 feet

No construction zone

Setback line

Minimum setback line

Baseline (natural or calculated dune line)

Houses

Revetment

Beach

Groin

Atlantic Ocean

10.3. Hypothetical projection of baselines, setback lines, and 20-foot "dune-protection" line (also the minimum setback line) in a standard erosion zone. Note the shoreline offset caused by the groin. The baseline for the natural shoreline to the left of the groin is the dune crest. The baseline for the armored beach (right of groin) is where the dune crest would have been under natural conditions. The distance between baseline and setback line is 100 feet, corresponding to 2.5 feet of erosion per year times 40 years.

back of the seawalls. The "ideal dune" creates a zone where a dune could form or be built if the structures were destroyed or removed.

Because shorelines around inlets are intrinsically unstable, their baselines are sometimes determined differently. Baselines in stabilized inlet zones (with groins, jetties, etc.) are determined in generally the same way as baselines in standard erosion zones. Shorelines in unstabilized inlet zones are likely to erode,

accrete, migrate, or otherwise change configuration, however, and baselines in these areas are determined to be the most landward position of that shoreline at any time in the last 40 years.

Once the baseline is established, the setback area is measured landward from it as 40 times the erosion rate. For example, if the erosion rate is 3 feet per year, the setback line is 120 feet from the baseline. Stable beaches have a minimum setback of 20 feet. To ensure protection of the landward half of the primary dune, this buffer zone exists 20 feet landward of all baselines.

Generally, all land seaward of the 20-foot line is in a "no-build" zone. Lands seaward of the setback line but landward of the 20-foot line are in a regulated zone. Most kinds of construction and other activity may be done in a setback area, but many things require a permit to ensure compliance with the law and the retreat policy. There are numerous exceptions and special conditions. Again, refer to appendix C for more information or seek professional assistance for specific situations.

Deed Disclosure

Of particular interest to prospective homeowners is the "deed disclosure" requirement of the Beachfront Management Act. Any sale or transfer of property affected by a setback line requires complete disclosure of the relationship of the property and buildings to the baseline and the setback line (fig. 10.3). Also required is the local erosion rate. The OCRM provides a form that can be completed to satisfy this disclosure requirement (see appendix B). The Beachfront Management Act says that "language reasonably calculated to call attention to the existence of baselines, setback lines, jurisdiction lines, and the erosion rate complies with this section." However, the law clearly stipulates that such information must be given in terms of the State Plane Coordinate System. Whether or not you use the OCRM form, you may need the assistance of the OCRM staff to provide the necessary information.

Every coastal property owner and resident should be aware of the essentials of this law, especially as it pertains to property and the local environment. The law is South Carolina's Coastal Zone Management Act: "Coastal Tidelands and Wetlands," Chapter 39 of Title 48 of the 1976 Code, as amended (July, 1990).

A Great Law Survives a Great Challenge

When the Beachfront Management Act became law in 1988, the Coastal Council was given two years to determine exact erosion rates and resolve certain other details, not the least of which was the legal basis for prohibiting building on private property. Another concern was the large size of setback areas on certain Unstabilized Inlet zones, most notably on Isle of Palms. The northeast end of Isle of Palms is generally accretional but is subject to

extreme episodic erosional events. Sediment bypassed across Dewees Inlet accumulates in offshore shoals, awaiting the right coastal conditions to move onshore. Where these shoals remain poised offshore they cause a natural wave refraction pattern that may cause extreme erosion both upcoast and downcoast of the shoal (see chapter 2).

Although this area is accretional in the long run, it is characterized by large erosional arcs, superimposed and overlapping throughout the inlet zone of the island (fig. 10.4). The most landward segments of these erosional shoreline arcs comprise "the most landward shoreline," the baseline defined by the Beachfront Management Act. (The baseline is the starting point for the setback measurement. The setback in this part of Isle of Palms is the minimum, 20 feet, because the area is accretional overall.) Many properties, even many developed properties, were affected by this regulation even though they were currently hundreds of feet from the shore. Some resolution or adjustment had to be found by July 1990.

In the middle of this area are two lots, then owned by Mr. David Lucas. The lots were of normal size, undeveloped, and separated by another lot. These properties had been completely submerged by the Atlantic Ocean four times in the 40-year period stipulated by the Beachfront Management Act (which speaks to the reason this methodology was employed). Mr. Lucas sued the State of South Carolina and the Coastal Council because the

Beachfront Management Act prevented him from making a profit on his land, at least during the period from 1988 to 1990.

If the regulations promulgated under the Beachfront Management Act should be found to constitute a taking of property without just compensation (a regulatory taking, or simply a "taking"), then that part of the law would be unconstitutional. The lawsuit created two major concerns. First, there was the possibility that the entire Beachfront Management Act would be overturned as unconstitutional. (So much for shoreline management based on our knowledge of where the shores will be in 40 years.) Second, if you can imagine a bad dream becoming a nightmare, there was a real possibility that if the Beachfront Management Act was overturned, any and all land-use regulations would begin to unravel. This would include everything from prohibitions against destruction of freshwater and saltwater wetlands to municipal zoning regulations. Given the enormous number of property owners who stood to profit from such deregulation, the *Lucas* case represented an almost unimaginable threat to South Carolina's environment. More than 120 "friends of the court" lined up on the two sides of the battle line.

The amended Beachfront Management Act of 1990 contains a section titled "Special Permits" that allows the OCRM to permit building in otherwise prohibited areas if these are not active beach or primary dunes. Mr. Lucas certainly qualified for a Special Permit. Indeed,

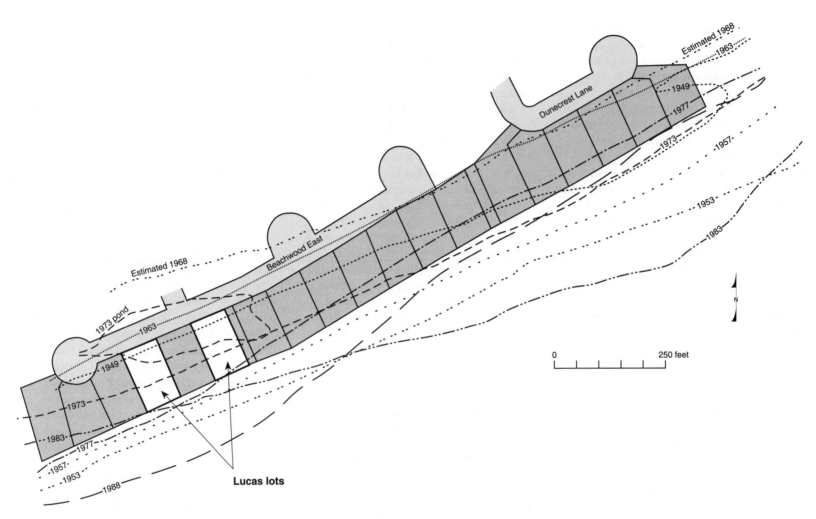

Estimated 1968
1963
Dunecrest Lane
1949
1977
1973
1957
1953
1983

Estimated 1968
Beachwood East
1973 pond
1963
1949
1973
1983
1977
1957
1953
1988

Lucas lots

0   250 feet

N

10.4. Description of historic shorelines along the northeast end of Isle of Palms showing the most landward shoreline (which equals the baseline in an unstabilized inlet zone). The lots owned by David Lucas are indicated at lower left. Modified after OCRM.

cases like his were the reason for the amendment. The entire State of South Carolina would have been better off if he had requested one; but the "property rights" movement had gained such momentum around this issue that in his own words he was "not allowed to." Without an application from him, the Coastal Council would not issue a permit, a position it came to regret.

The South Carolina Supreme Court heard the case of *Lucas v. State of South Carolina* in 1991 and ruled that no taking had occurred. In 1992 the Supreme Court of the United States reversed this ruling, saying that by regulating the setback area, South Carolina had in effect taken Mr. Lucas's property. Two significant questions that arose during this hearing were, to Mr. Lucas: "Why didn't you simply apply for a [Special] Permit?" and to attorneys for the Coastal Council: "Why didn't you simply issue him a permit?"

The case was remanded to the South Carolina Supreme Court to determine the appropriate settlement. The Lucas properties were purchased by the state, and at the time of this writing one has been resold and developed. The settlement was approximately $1.675 million for the lots and the loss of the value of the lots from 1988 to 1992.

The *Lucas* case technically ends there. But the story has taken on a life of its own. Environmental and real estate lawyers point to this case as one of the most misquoted and misunderstood precedents in recent history. Granted,

the *Lucas* case is one more story of a developer's windfall at the taxpayers' expense. But other than that, nothing happened! The fears of environmentalists and scientists did not materialize. The U.S. Supreme Court did not issue a grand, sweeping decision to bolster the property rights movement. It simply said that the South Carolina Supreme Court erred in its 1991 ruling. The basis for the state's land-use laws was not rendered unconstitutional. In fact, the Beachfront Management Act itself remains unchanged as a result of this challenge. The U.S. Supreme Court in effect clearly established the constitutionality of the law.

At the time of this writing the "Lucas properties" are once again highly erosional. Dune scarps more than 8 feet tall characterize this segment of shoreline. Hundreds of truckloads of sand have been hauled to the site. Bulldozers scrape the low-tide beach daily to replace the quickly eroding sand, and hopefully encourage the shoal attachment. Wide disagreement exists between property owners, island residents, the Isle of Palms government, and the OCRM as to how or even whether to solve the erosion problem. Sandbag revetments have been installed, but property owners are demanding more permanent relief. Apparently, nature wasn't paying attention to the importance of this landmark case.

The Future of Coastal Zone Regulation in South Carolina

The Beachfront Management Act has been a definite step forward. Some totally imprudent development is being controlled because of it, and new development is likely to have a longer property life and higher value. Migrating dunes and beaches that would have been destroyed a few years ago are now being preserved for the benefit of all. The basis of our coastal management is our understanding of where our shorelines will be in 40 years.

One area that still generates concern is the OCRM itself. As in any type of land-use control, the strength of the regulations that govern coastal development is determined by how those responsible for enforcement interpret them. In the past, the Coastal Council was responsible for many decisions that saved buildings but severely degraded recreational beaches. This is the legacy of seawalls and the power of private property. There is a great need in South Carolina for the OCRM to take a long view, in terms of both time and distance, that will establish a clear priority to preserve South Carolina's beaches for future generations.

If you are still uncertain about the need for coastal regulation, consider the mandate of the OCRM. That agency has the double task of preserving South Carolina's beaches for all her citizens and employing the best solutions for coping with the erosion of private property—

in that order. Truly, the twin goals of this mission are at cross purposes. In any case, it is the OCRM's job to manage a precious resource that, in spite of the best efforts by the best professionals, is growing smaller through erosion and development every day, every year. All that can be done now is to manage the problem and minimize the inevitable loss of South Carolina's beaches and property. This goal cannot be achieved without the support of the citizens, who own the resource. If you would like for your grandchildren to be able to enjoy the beaches of South Carolina, your support of the mission of the OCRM is the logical minimum requirement.

Water Quality and Waste Disposal

Protecting the water resources of barrier islands as well as the mainland coast is essential to safeguarding the various uses of the coast. Fisheries, all forms of water recreation, and the general ecosystem depend on high-quality surface waters. Potable water is drawn mainly from groundwater, which must also be of high quality. As noted in chapters 5 and 6, our water resources are under great stress, and the existing pollution is costly to both local communities and the state. When shellfishing waters are closed or the health of local residents is threatened, the loss is more than economic.

The Federal Water Pollution Control Act Amendments of 1972 (P.L. 92-500) control any type of land use that generates, or may

generate, water pollution. The dredging and filling of wetlands and water bodies is regulated through the U.S. Army Corps of Engineers (see appendix B, "Dredging, Filling, and Construction in Coastal Waterways"; "Sanitation and Septic System Permits"; and "Water Resources"). The Marine Protection, Research and Sanctuaries Act of 1972 (P.L. 92-532) regulates dumping into ocean water. The Water Resources Development Act of 1974 (P.L. 93-251) provides for comprehensive coastal zone planning.

The South Carolina Ground Water Use Act of 1969 authorizes the South Carolina Water Resources Commission to establish rules and regulations to manage groundwater use. This agency is now part of the Department of Health and Environmental Control, which is also responsible for controlling and preventing pollution of groundwater. The DHEC is the agency responsible to the U.S. Environmental Protection Agency for administering section 208 and P.L. 92-500 in the development of a state waste treatment management plan (appendix D, ref. 117). The DHEC's jurisdiction over the quality of the state's air and water is defined under Title 48 of the South Carolina Pollution Control Act. Surface water quality standards were established by the DHEC in the 1981 Water Classification Standards System. More recently, the DHEC administers the Section 404 Wetlands Certification Process.

Before drilling a well or installing a septic system, check with the county or city administrator's office or the health department

to find out about necessary permits. In the case of septic tanks, the local county has jurisdiction if the proposed project involves 45 lots or less.

Building Codes

Most progressive communities require new construction to adhere to the provisions of a recognized building code. If you plan to build in an area that does not follow such a code, you would be wise to insist that your builder do so anyway to meet your requirements. Local building officials in storm-prone areas often adopt national codes that contain building requirements for protection against high wind and water. Compiled by knowledgeable engineers, politicians, and architects, these codes regulate the design and construction of buildings and the quality of building materials. Examples of such codes are the Standard Building Code, used chiefly on the southeast and Gulf coasts, and the Uniform Building Code, used mainly on the West Coast. Both are excellent codes (see appendix D, "Coastal Construction").

The purpose of these codes is to provide minimum standards to safeguard lives, health, and property. The codes protect you from yourself as well as from your neighbor. Parts of poorly constructed buildings, or even an entire building, on your property might become dislodged during a storm and damage neighboring property.

South Carolina has neither a general state code (other than for state-owned buildings) nor a specific coastal building code; however, the state is beginning to examine the building codes of other coastal states. In the future, South Carolina may adopt such a code to strengthen building requirements in areas not covered by a local code. Before building, check with the local city or county administrator's office about local building code requirements.

Mobile Home Regulations

Mobile homes differ in construction and anchorage from "permanent" structures. The design, shape, lightweight construction materials, and other characteristics required for mobility and for staying within axle weight limits create a unique set of potential problems for residents of these dwellings. Because of their thinner walls, for example, mobile homes are more vulnerable to wind and windborne projectiles than are permanent homes. Some coastal states have separate building codes or requirements for mobile homes; South Carolina, however, does not. Many counties and municipalities within South Carolina do have such ordinances, and the owner of a mobile home should check with the appropriate authority.

Mobile home anchorage is commonly regulated by local ordinance. Tie-downs should be required to make the structure more stable against wind stress (see chapter 8 for recommendations). Violations of anchorage or foundation regulations may go undetected unless there is a sufficient number of conscientious inspectors to monitor trailer courts. One poorly anchored mobile home can severely damage adjacent homes whose owners followed sound construction practice. Some operators or managers of mobile home parks are alert to such problems and see that they are corrected; others simply collect the rent.

The spacing of mobile homes also may be regulated by local ordinance. By requiring residents to leave some open space between homes, this type of ordinance preserves aesthetic value for the neighborhood and helps to maintain a healthier environment. For example, if mobile home septic tanks are too close together, there is the potential for groundwater or surface water pollution. Similarly, if mobile homes are placed too close to finger canals, the canal water may become polluted. Check with your city or county administrator's office about mobile home regulations.

Hazards, Economics, and Politics

Our survey of South Carolina's coastal communities reveals a great diversity of hazards, attitudes, and responses. Nevertheless, enough similarities exist to make a few generalizations.

1. Town sites generally came into existence without hazard planning. Plantations, port facilities, church camps, hunting clubs, and the resorts they ultimately became evolved haphazardly (with a few exceptions, such as Kiawah Island). Sites on barrier island shores, along estuaries, on deltas and floodplains, and in seismic zones were often the most hazardous. Barrier island towns in particular were platted in traditional grids over fragile, dynamic environments rather than being developed with site stability, suitability, and low risk in mind.

2. Where early residents learned from experience, low-risk sites tended to be developed first, leaving high-risk sites and areas to accommodate later growth (e.g., Pawleys Island). This later development, when threatened, has usually opted for engineering "solutions" rather than relocation. These projects are increasingly expensive and often ineffective.

3. Politicians and the political pressures to which they react are oriented toward giving priority to economic development and management, *not* toward protecting the electorate. Development is seen as progress, and, unchanged since pioneer days, such progress is still considered our manifest destiny.

4. "Protective" regulations to reduce natural hazards (e.g., prohibitions on fill and development of wetlands) may be threatening to developers and some property owners. Developers often resist regulations designed to protect property owners.

5. Politicians are usually drawn from the economic community. They are often them-

selves owners of undeveloped acreage, developers, suppliers of materials, lawyers, realtors, businesspeople, and other professionals who benefit from growth and development. Even if no conflict of interest is intended, they have a stake in development to protect. So the approval of a new development may be influenced by a board member's vision of lumber sales or restaurant patrons rather than an evaluation of the development's hazard risk potential.

6. Politicians are the employers, while the day-to-day work is carried out by the employees: the hired town manager, planner, and community development personnel. These employees by and large do an excellent job for South Carolina communities. They are knowledgeable, realistic, and committed public servants, but they answer to the elected politicians, not to the general public.

7. When disaster does strike, we depend on firefighters and police as our first line of defense. The people with the greatest responsibility for public safety sometimes seem to be the least appreciated; many communities rely on volunteers rather than a salaried fire-fighting staff, and media comments sometimes suggest that police are overpaid (the high salaries resulting from overtime necessitated by understaffing). The dedication and fearlessness of these brave people during hazardous and extremely stressful times often means the difference between life and death for hundreds or thousands affected by natural disasters.

8. Collective community attitudes are variable. For example, Isle of Palms and Sullivans Island have a high percentage of permanent residents. Their perceived planning needs demonstrate a higher level of responsibility than more transiently populated communities such as Harbor Island or Garden City. Newcomers in the latter communities have had no experience with the surprises of shoreline or coastal living. Their attitude is different with respect to planning, and they are often more likely to locate in newer, higher-risk developments.

9. Developers are in business to make money, not to protect the public. The emphasis is on build and sell, not on analyzing site-specific hazard risk.

10. Banks and other lenders do have a stake in property mortgages. In some cases, lenders can be a source of information on risk; however, if your credit is good, you can get the loan in spite of the risk. You will, however, be required to have federal flood insurance.

11. Catastrophes create several responses that affect both the local community and communities at large. Postcatastrophe "recovery" is often a time of shock and haste to put things right again, rather than a time of careful relocation and risk reduction. Houses and multifamily housing units are rebuilt "bigger and better" in the same high-risk zones. Big catastrophes bring new and/or stronger regulations; for example, higher insurance rates, upgraded building codes, prohibitions or restrictions on future development, and mitigation against re-

currence of the hazard. Expect them.

12. The levels of South Carolina's management and politics are as diverse as the communities and hazards themselves. There are towns, cities, counties, state, and federal regulatory agencies. There are also special regulations for private and public lands such as units of The Nature Conservancy, Heritage Trust, COBIA, wildlife preserves, and other restricted lands.

The individual citizen and property owner is the final decision maker. The foregoing chapters provide a starting point to identify and evaluate hazards in terms of potential risk to both property and human health and safety. The appendixes that follow contain hazard checklists, contact agencies, and recommended reading to continue to broaden your South Carolina horizons.

Coastal Regulation: Yesterday, Today, and Tomorrow

The recognition of the need for coastal zone management is as old as human occupation and development within the coastal high-hazard zone. The consequences of unfettered development and overutilization of resources that are of greater value when left undisturbed demonstrate the need for management. Coastal management was and is influenced by changing styles of government and politics, however; as our knowledge and experience of coastal dynamics and change increase, so do the levels of regulation. Proper management of South Carolina's coastal zone is vital not just to natural environments and resources but also to the state's economy, the welfare and safety of its population, and the investment in its coast.

New and existing development in the coastal zone must meet or exceed the minimum requirements of prudent management. Whether you are a developer or an individual planning to live in the coastal zone, you must seek the necessary permits. Expect stricter enforcement of existing regulations and new, "tighter" regulations in the years to come as more and more people choose to live with South Carolina's coast.

An Unending Game: Only the Players Change

Earthquake or calm, hurricane or sun, eroding shore or accreting land, storm-surge flood or migrating dune, win or lose, the gamble of coastal development will continue. If you choose your site with natural safety in mind, follow structural engineering design in construction, and take a generally prudent approach to living at the shore, then you become the gambler who knows when to hold them, when to fold them, and when to walk away.

Our goal is to provide guidance for today's and tomorrow's players. This book is not the last, or by any means the most complete, guide to coastal living, but it does provide a starting point. The appendixes that follow list additional resources that we hope every reader will pursue.

Natural disasters such as those described in this book can strike at any time, often without warning. Even with modern hurricane and flood prediction and warning capabilities, so many uncertainties remain that it is impossible to make perfect watch, warning, and evacuation calls every time. For example, Hurricane Hugo's forward speed and absolute strength grew so rapidly during the few short hours before it made landfall that the coast from Virginia to Florida was still preparing for a category 3 storm when Hugo rammed the Charleston area as a category 4.

With respect to the uncertainties involved in preparing for natural disasters, we include in this appendix several safety checklists with guidelines to preparing for, riding out, and recovering from natural disasters. Some precautions and preparations are the same for all disasters, and these are included as a generic "Disaster Preparation Guide." Specifics for hurricanes, floods, and earthquakes follow the general information. Keep these checklists handy, and use them to protect your family and property.

Also listed are several publications you may want to include in your home library. In most cases the publications are free and readily available with just a phone call or postcard.

General Disaster Preparation Guide

Find out what natural hazards are likely to affect your community (or vacation rental site).

Are earthquakes possible? Are they likely? Are you located on a floodplain, and have heavy rains been predicted or recently occurred? Have you just moved your entire family into your dream vacation home right on the oceanfront while a tropical disturbance is forming or growing out to sea?

One of the best all-around sources of information on disaster preparedness is a book called *Are You Ready?* by FEMA (revised September 1993). This 86-page publication is extensive in its coverage of natural and technological (nuclear, toxic spills) disasters. Write to FEMA at the following address and ask for publication H-34.

FEMA
Attn: Publications
P.O. Box 70274
Washington, DC 20024

You should have a disaster supply kit already put together with the essentials needed for any emergency from power outages to hurricanes.
—Stock adequate supplies:
  —transistor (battery-powered) radio
  —weather radio with alarm function for direct National Weather Service broadcasts
  —fresh batteries
  —hammer
  —boards (for securing windows and doors for hurricanes)
  —pliers
  —hunting knife
  —tape
  —first-aid kit
  —prescribed medicines
  —candles
  —matches
  —nails
  —ax (to cut an emergency escape opening if you go to the upper floor or attic of your home)
  —rope (for escape to the ground when water subsides)
  —plastic drop cloths, waterproof bags, and ties
  —containers of water
  —canned food, juices, soft drinks
  —enough food for at least three days and enough water for more than three days; select food that does not require cooking or refrigeration
  —water purification tablets
  —insect repellent
  —gum, candy
  —life jackets
  —charcoal bucket, charcoal, and charcoal lighter
  —buckets of sand
  —disinfectant
  —flashlights
  —hard-top headgear
  —fire extinguisher
  —can opener and utensils (knives, forks, spoons, cups)
Make sure you know how to shut off electricity, gas, and water at main switches and

valves. Know where the gas pilots are and how the heating system works. Be ready and able to secure all loose items before you leave so that your property won't cause harm to others and you will be able to reenter your home safely.

Make a record of your personal property. Take photographs of or videotape your belongings and store the record in a safe place.

Keep insurance policies, deeds, property records, and other important papers in a safe place away from your home.

Hurricane Safety Checklist

When (or Before) a Hurricane Threatens

—Most important, know the *official evacuation route* for your area. You will not be asked to evacuate unless your life is in danger, so *evacuate as directed* by local emergency preparedness officials.
—Many communities contain information about hurricane preparedness and evacuation in the local telephone book. Check and see if the information is there and make sure that everyone in the house is familiar with it.
—Read your newspaper and listen to radio and television for official weather reports and announcements.
—Secure a reentry permit if necessary. Some communities allow only property owners and residents with proper identification or tags to return in the storm's immediate wake.

—Pregnant women, the ill, and the infirm should call a physician for advice.
—Be prepared to turn off gas, water, and electricity where it enters your home.
—Make sure your car's gas tank is full.
—Secure your boat. Use long lines to allow for rising water.
—Secure movable objects on your property:
 —doors and gates
 —outdoor furniture
 —shutters
 —garden tools
 —hoses
 —garbage cans
 —bicycles or large sports equipment
 —barbecues or grills
 —other
—Board up or tape windows and glazed areas. Close storm shutters. Draw drapes and window blinds across windows and glass doors, and remove furniture in their vicinity.
—Check mobile home tie-downs.
—Your primary line of defense is *early evacuation*. If you are unable to evacuate, you should also do the following.
—Know the location of the nearest emergency shelter. Go there if directed by emergency preparedness officials.
—When a flood or hurricane is imminent, fill tubs and containers with water enough for one week (a minimum of one quart per person per day).

Special Precautions for Apartments/ Condominiums

—Designate one person as the building captain to supervise storm preparations.
—Know your exits.
—Count stairs on exits; you may be evacuating in darkness.
—Locate safest areas for occupants to congregate.
—Close, lock, and tape windows.
—Remove loose items from terraces (and from your absent neighbor's terraces).
—Remove (or tie down) loose objects from balconies or porches.
—Assume other trapped people may wish to use the building for shelter.

Special Precautions for Mobile Homes

—Pack breakables in padded cartons and place on the floor.
—Remove bulbs, lamps, and mirrors, and put them in the bathtub.
—Tape windows.
—Turn off water, propane gas, electricity.
—Disconnect sewer and water lines.
—Remove awnings.
—*Leave.* Don't stay inside for *any* reason unless you can absolutely ascertain that the hazards from rising floodwaters are greater than the danger from wind.

Special Precautions for Businesses

—Take photographs of building and merchandise before the storm.
—Assemble insurance policies.
—Move merchandise away from plate glass.
—Move merchandise to the highest locations possible.
—Cover merchandise with tarps or plastic.
—Remove outside display racks and loose signs.
—Take out file drawers, wrap them in trash bags, and store in high places.
—Sandbag spaces that may leak.
—Take special precautions with reactive or toxic chemicals.

If You Remain at Home

—*Never* remain in a mobile home; seek official shelter.
—Stay indoors. Remember, the first calm may be the hurricane's eye. Do not attempt to change your location during the eye unless absolutely necessary. Remain indoors until an official all-clear is given.
—Stay on the *downwind* side of the house. Change your position as the wind direction changes.
—If your house has an inside room (away from all outdoor walls), it may be the most secure part of the structure. Stay there.
—Keep a continuous watch for *official* information on radio and television.

—Keep calm. Your ability to meet emergencies will help others.

If Evacuation Is Advised

—Leave as soon as you can. Follow official instructions only. Ignore rumors.
—Follow predesignated evacuation routes unless those in authority direct you to do otherwise.
—Take these supplies:
  —reentry permit
  —change of warm, protective clothes
  —first-aid kit
  —baby formula
  —identification tags: include name, address, next of kin (wear them)
  —flashlight
  —food, water, gum, candy
  —rope, hunting knife
  —waterproof bags and ties
  —can opener and eating utensils
  —disposable diapers
  —special medicine
  —blankets and pillows, in waterproof casings
  —transistor (or battery-powered) radio
  —fresh batteries (for radio *and* flashlight)
  —bottled water
  —purse, wallet, valuables
  —life jackets
  —games and amusements for children
—Disconnect all electric appliances except refrigerator and freezer; their controls should

be turned to the coldest setting and the doors kept closed.
—Leave food and water for pets. Seeing-eye dogs are the only animals allowed in shelters.
—Shut off water at the main valve (where it enters your home).
—Lock windows and doors.
—Keep important papers with you:
—driver's license and other identification
—insurance policies
—property inventory
—Medic Alert or other advice to convey special medical information

During the Hurricane

—Stay indoors and away from windows and glassed areas.
—If you are advised to evacuate, *do so at once.*
—Listen for weather bulletins and official reports.
—Use your telephone only in an emergency.
—Follow official instructions only. Ignore rumors.
—Keep a window or door *open* on the side of the house opposite the storm winds.
—Beware the eye of the hurricane. A lull in the winds does not necessarily mean that the storm has passed. Remain indoors unless emergency repairs are necessary. Be cautious. Winds may resume suddenly, in the opposite direction and with greater force than before. Remember, if wind direction

does change, the open window or door must be changed accordingly.

—Be alert for rising water. Stand on furniture if necessary.

—If electric service is interrupted, note the time. Take the following steps when the electricity goes out:

—Turn off major appliances, especially air conditioners.

—Do not disconnect refrigerators or freezers. Their controls should have been turned to the coldest setting and the doors kept closed to preserve food for as long as possible.

—Keep away from fallen wires. Report the location of such wires to the utility company.

If You Detect Gas

—Do not light matches or cigarette lighters or turn on electrical equipment, not even a flashlight.

—Extinguish all flames.

—Shut off gas supply at the meter. Gas should be turned back on only by a gas service professional or licensed plumber.

Water

—The only *safe* water is the water you stored before it had a chance to come in contact with floodwaters.

—Should you require more water, be sure to boil it for 30 minutes before using it.

—If you are unable to boil water, treat it with water purification tablets. These are available at camping stores.

Note: An official announcement will proclaim tap water safe. Boil or treat all water except stored water until you hear the announcement.

After the Hurricane Has Passed

—Listen for official word that the danger has passed. Don't return to your home until officially directed to do so.

—Watch out for loose or hanging power lines as well as gas leaks. People have survived storms only to be electrocuted or burned. Fire protection may be nonexistent because broken power lines and fallen debris in the streets are blocking access.

—Walk or drive carefully through storm-damaged areas. Streets will be dangerous because of debris, undermining by washout, and weakened bridges. Watch out for animals that may act irrationally after being driven out by floodwaters.

—Looting may be a problem in certain areas. Police protection may be nonexistent. Do not participate in such illegal acts and do not try to stop others. Wait for police or National Guard or other officials to arrive.

—Eat nothing and drink nothing that has been touched by floodwaters.

—Place spoiled food in plastic bags and tie them securely.

—Dispose of all mattresses, pillows, and cushions that have been in floodwaters.

—Contact relatives as soon as possible.

—If you use an electric generator for home power, make sure your house's main circuit breaker switch is off. This will prevent your home-generated electricity from "leaking" out to the main power lines. After Hurricane Hugo in 1989, several power line repairers were electrocuted by electricity from the home generators of thoughtless individuals. Save a life, make sure your main circuit breaker is off!

Note: If you are stranded, signal for help by waving a flashlight at night or a white cloth during the day.

Riverine and Flash Flood Safety Checklist

What to Do Before a Flood

—Know the terminology

—Flood watch:

—flooding is possible

—stay tuned to NOAA weather radio or commercial radio or television

—Flash flood watch:

—flash flooding, which can result in raging waters in just a few minutes, is possible

—move to higher ground, a flash flood could occur without any warning

—stay tuned to radio or television

—Flood warning:

—flooding is occurring or will occur soon

—if evacuation is advised, do so immediately
—Flash flood warning:
  —a flash flood is occurring
  —seek higher ground on foot immediately
—Find out from your local emergency management office whether your property is in a flood-prone area. Learn the elevation of your area. Learn about the likely flooding scenario of your lot/neighborhood/community.
—Identify dams in your area and determine whether they pose a hazard.
—Purchase flood insurance (flood losses are *not covered* under homeowner's insurance policies); it is widely available through the NFIP.
—Prepare a family plan:
  —have a portable radio, flashlight, and emergency supplies
  —be prepared to evacuate
  —learn local evacuation routes
  —choose a safe area in advance
  —plan a family meeting place in case you are separated and cannot return home

During a Flood

—Flood watch (2–3 days for flood; 2–12 hours for flash flood)
—*If you have time,* bring outdoor garden equipment and lawn furniture inside or tie it down.

—*If you have time,* move furniture and other items to higher levels (for flood).
—Fill you car's gas tank (for flood).
—Listen to radio or TV for up-to-the-minute information.
—Flood warning (24–48 hours for flood; 0–1 hour for flash flood)
—Evacuate, if necessary, when a flood warning indicates, and follow instructions.
—Do not walk or drive through floodwaters.
—Stay off bridges covered by water.
—Heed barricades blocking roads.
—Keep away from waterways during heavy rain; if you are driving in a canyon area and hear a warning, get out of your car and get to high ground immediately.
—Keep out of storm drains and irrigation ditches.
—Listen to radio or TV for up-to-the-minute information.

After a Flood

—Stay away from floodwaters, they could be contaminated.
—Stay away from moving water.

Earthquake Safety Checklist

Before an Earthquake

—Have a plan of action, and know what to do afterward.

—Have a family reunion plan.
—Have an out-of-state family contact.
—Have supplies on hand such as water, a flashlight, a portable radio, food, a fire extinguisher, and tools.
—Bolt down bookshelves and water heaters, secure cabinets.

During an Earthquake

—Get under a heavy table or desk and hold on, or sit or stand against an inside wall.
—Keep away from windows.
—If you are indoors, stay indoors.
—If you are outdoors, stay outdoors away from falling debris, trees, and power lines.
—If you are in a car, stay in the car.
—Don't use elevators.

After an Earthquake

—Expect aftershocks.
—Check gas, water, and electrical lines and appliances for damage.
—Use a flashlight (not a match!) to inspect for damage.
—Turn off main gas and electricity valves.
—Don't venture into damaged areas.
—Don't use telephones except in emergencies.
—Don't use vehicles except in emergencies.
—Use a portable radio for information.

Numerous agencies at all levels of government are engaged in planning, regulating, and studying coastal development in South Carolina. Perhaps the most important state agency is the Office of Ocean and Coastal Resource Management (OCRM) of the South Carolina Department of Health and Environmental Control (DHEC). The OCRM issues permits for various phases of construction and provide information on development to the homeowner, developer, or planner. Following is an alphabetical list of topics related to coastal development; under each topic are the names of agencies to consult for information on that topic. Some of these sources also provide information about noncoastal areas.

Among the references listed in appendix D are some that provide additional agency listings and basic information of interest to the coastal dweller. In particular, persons needing a more complete list of federal and state agencies involved in coastal development should obtain a copy of *Natural Hazard Management in Coastal Areas* (appendix D, ref. 100; or contact the OCRM), a comprehensive report and atlas for South Carolina and Georgia.

Aerial Photography and Remote-Sensing Imagery

Persons interested in aerial photography, remote-sensing imagery, or agencies that supply aerial photographs or images should contact the appropriate offices listed below.

For historic listing of available photography (type, scale, year flown, coverage, percentage of cloud cover, etc.) contact:

Earth Science Information Center
U.S. Geological Survey
12201 Sunrise Valley Drive
Reston, VA 22092
Phone: (703) 860-6045

Request "APSRS Aerial Photography Summary Record System: No. 23/North Carolina, South Carolina" and accompanying microfiche.

Recent aerial photography should be available from:

U.S. Department of Agriculture, Agricultural Stabilization and Conservation Service
Aerial Photography Field Office
2222 West, 2300 South
P.O. Box 30010
Salt Lake City, UT 84130-0010
Phone: (801) 975-3503

Request "Status of Aerial Photography Coverage" for South Carolina. Black-and-white vertical aerial photos are available for the coastal counties.

Offices that may have aerial photographs available for inspection but are generally not suppliers include the following:
South Carolina Department of Health and Environmental Control
Office of Ocean and Coastal Resource Management
4130 Faber Place, Suite 300
Charleston, SC 29405
Phone: (803) 744-5838

South Carolina Natural Resources Department
Marine Resources Division
P.O. Box 12559
Charleston, SC 29412
Phone: (803) 762-5000

Other, more conveniently located sources that may have aerial photographs of your area of interest available for inspection include the office of the tax assessor in your county, departments of geology or geography at local colleges and universities, the district office of the U.S. Army Corps of Engineers, and the Division of Geology, South Carolina State Development Board, in Columbia. For information on satellite imagery contact:

EROS Data Center
U.S. Geological Survey
Sioux Falls, SD 57198
Phone: General information (605) 594-6511
Orders (605) 594-6151

Archives and Records

The following is a source of historic information on coastal counties and a possible source of historic maps and photographs:

Director
Department of Archives and History
1430 Senate Street
P.O. Box 11669
Columbia, SC 29201
Phone: (803) 734-8577

Beach Erosion

Information on beach erosion, inlet migration, floods, and high winds is available from the following:

South Carolina Department of Health and Environmental Control
Office of Ocean and Coastal Resource Management
4130 Faber Place, Suite 300
Charleston, SC 29405
Phone: (803) 744-5838

District Engineer
U.S. Army Corps of Engineers
P.O. Box 919
Charleston, SC 29402-0919
Phone: (803) 727-4236

South Carolina Natural Resources Department
Marine Resources Division
P.O. Box 12559
Charleston, SC 29412
Phone: (803) 762-5000

South Carolina Sea Grant Consortium
287 Meeting Street

Charleston, SC 29401
Phone: (803) 727-2078

Bridges and Causeways

The U.S. Coast Guard has jurisdiction over permits to build bridges or causeways that will affect navigable waters. Information is available from:

Commander
7th Coast Guard District
909 SE 1st Avenue
Miami, FL 33130
Phone: (305) 536-4108

Building Codes and Zoning

South Carolina does not have a state building code. Counties or municipalities are responsible for such codes and their enforcement. The old Southern Building Code, now called the Standard Building Code, is the basis for most local codes. Some municipalities add more stringent requirements and may have special regulations for mobile homes. Communities participating in the National Flood Insurance Program will have building elevation requirements in order to meet the specifications of the program. For the specific code in your area, contact the city or county administrator.

Check to be sure that the property in which you are interested is zoned for your intended use and that adjacent property zones do not

conflict with your plans. For information, contact the city or county building inspector.

Coastal Zone Management

Act 123 of the 1977 South Carolina General Assembly established the South Carolina Coastal Council and charged it "to promote economic and social improvement of the citizens of the State and to encourage development of coastal resources in order to achieve such improvement with due consideration for the environment and within the framework of a coastal planning program that is designed to protect the sensitive and fragile areas from inappropriate development and provide adequate environmental safeguards with respect to the construction of facilities in the critical areas of the coastal zone" (section 2 [B][1]). This plan for a state coastal management program was approved by the South Carolina General Assembly and Governor Richard W. Riley on February 14, 1979.

The Coastal Council, since renamed the Office of Ocean and Coastal Resource Management, implements a two-tiered coastal management program. Its direct permitting authority (first tier) extends over activities proposed on or in any of four "critical areas" of the coastal zone: coastal waters, tidelands, beaches, and primary oceanfront sand dunes. Many of these activities are also issued permits by the U.S. Army Corps of Engineers. In such cases the Corps and the OCRM issue a joint public notice for the activity to expedite the permitting process.

The OCRM also exercises an indirect, or "certification," authority (second tier) over all activities within the eight-county coastal zone that require a permit or permits from other state agencies (e.g., domestic wastewater permits, water supply permits, air permits, and sanitary landfill permits are all issued by the DHEC, and well-water permits are issued by the South Carolina Water Resources Commission). If the OCRM finds the proposed activity inconsistent with coastal management program policies, the activity cannot be permitted by the other agency or agencies. The OCRM strongly urges developers to contact its office at the preliminary planning stages of development. This early coordination will advise developers of OCRM policies regarding that particular type of activity and will result in wise land-use practices consistent with the coastal management program. Early coordination results in a smooth process of permit issuance and certification and often can save developers much time and unnecessary expense. For further information, see chapter 10 or contact:

South Carolina Department of Health and Environmental Control
Office of Ocean and Coastal Resource Management
4130 Faber Place, Suite 300
Charleston, SC 29405
Phone: (803) 744-5838

Consultants

It is inappropriate for the authors of this book to endorse any individual or firm as a recommended coastal or construction consultant. We encourage prospective buyers and property owners to seek expert advice on housing construction safety and site safety with respect to coastal hazards. The offices listed in this appendix under other topics and the offices of your local government are sources of advice regarding appropriate private consultants for your particular problem. In addition, state colleges and universities, particularly those with coastal geologists and coastal engineers such as the University of South Carolina, Clemson University, and the College of Charleston, or conservation organizations such as the National Audubon Society, may know of reputable consultants for referral.

Disaster Assistance

Emergency Preparedness Division
1429 Senate Street
Columbia, SC 29201
Phone: (803) 734-8020

Dredging, Filling, and Construction in Coastal Waterways

Section 13 (C) of the South Carolina Coastal Zone Management Act (Act 123) of 1977, as amended by the Beachfront Management Act of 1990, states that "no person shall fill, remove, dredge, drain or erect any structure on or in any way alter any critical area without first obtaining a permit from the Council [the OCRM]." For application forms and instructions contact:

South Carolina Department of Health and Environmental Control
Office of Ocean and Coastal Resource Management
4130 Faber Place, Suite 300
Charleston, SC 29405
Phone: (803) 744-5838

Federal law requires any person who wishes to dredge, fill or place any structure in navigable water (almost any body of water) to apply for a permit from the U.S. Army Corps of Engineers. Information is available from:

Operations Division
Charleston District, U.S. Army Corps of Engineers
P.O. Box 919
Charleston, SC 29402
Phone: (803) 727-4330

The American Shore and Beach Preservation Association publishes a quarterly journal, *Shore and Beach,* which features papers concerned with dredging and construction along our nation's coasts. For more information, write:
American Shore and Beach Preservation Association

412 O'Brien Hall
University of California
Berkeley, CA 94720

Dune Alteration

The OCRM has permit authority over the beaches and primary oceanfront sand dunes. Disturbance of these dunes is prohibited. Even special permits (see chapter 10) are prohibited if the activity would affect an active beach or primary dune. Some local communities have regulations even stricter than the state law.

According to the Beachfront Management Act (30-13 [B][1]), the construction of walkways over sand dunes does not require a permit provided the walkways meet the following requirements: "a) be constructed of wood; b) have a maximum width of six feet; c) conform with the contour of the dunes; d) not displace any sand in a critical area; e) be constructed with as little environmental damage as possible."

Before considering any construction that may alter a sand dune critical area, you should contact the OCRM to see if a permit is required. Penalties are imposed for violating the state's Coastal Zone Management Act. Call or write:

South Carolina Department of Health and Environment Control
Office of Ocean and Coastal Resource Management
4130 Faber Place, Suite 300

Charleston, SC 29405
Phone: (803) 744-5838

Environmental Affairs

Environmental Protection Agency
Washington, DC 20460
Phone: (202) 260-2090
EPA Region 4 Office
345 Courtland Street, NE
Atlanta, GA 30365
Phone: (404) 347-3004

South Carolina Department of Health and Environmental Control
2600 Bull Street
Columbia, SC 29201
Phone: (803) 734-5000

Flood Insurance (see *Insurance*)

Geologic Information

Earth Science Information Center
U.S. Geological Survey
12201 Sunrise Valley Drive
Reston, VA 22092
Phone: (703) 860-6045

Request geologic and water supply reports and maps for South Carolina; free index.

U.S. Geological Survey
Water Resources Division
720 Gracern Road, Suite 129
Columbia, SC 29201-7651
Phone: (803) 750-6100

State Geologist
South Carolina Geological Survey
5 Geology Road
Columbia, SC 29210
Phone: (803) 896-7708

University of South Carolina
Department of Geological Sciences
Columbia, SC 29208
Phone: (803) 777-3120

Hazards (see also *Beach Erosion* and *Insurance*)

Literature describing natural hazards on barrier islands is available from:

South Carolina Sea Grant Consortium
287 Meeting Street
Charleston, SC 29401
Phone: (803) 727-2078

Request a list of publications.

UNC Sea Grant Program
Box 8605
North Carolina State University
Raleigh, NC 27695-8605
Phone: (919) 515-2454

Request a list of publications.

National Oceanic and Atmospheric Administration
Office of Ocean and Coastal Resources Management
1305 E-W Highway
Silver Spring, MD 20910
Phone: (301) 713-3115

Information regarding earthquake hazards is available from:

Earthquake Program Manager
South Carolina Emergency Preparedness Division
1429 Senate Street
Columbia, SC 29201
Phone: (803) 734-8020

Earthquake Education Center
Charleston Southern University
P.O. Box 118087
Charleston, SC 29423-8087
Phone: (803) 863-8090

Health (see also *Sanitation and Septic System Permits*)

The local health department in your city or county will provide information on home waste-treatment systems, water supply systems, and similar health matters. Questions relating to water quality also may be directed to:

South Carolina Department of Health and Environmental Control 2600 Bull Street

Columbia, SC 29201
Phone: (803) 734-5000

History

South Carolina Department of Archives and History
1430 Senate Street
Columbia, SC 29201
Phone: (803) 734-8577

South Carolina Library
University of South Carolina
Columbia, SC 29208
South Carolina Historical Society
100 Meeting Street
Charleston, SC 29401
Phone: (803) 723-3225

Housing (see *Subdivisions*)

Hurricane Information

National Hurricane Center
11691 SW 17th Street
Miami, FL 33165-2149
Phone: (305) 229-4470

National Oceanic and Atmospheric Administration
Office of Ocean Coastal Resources Management
1305 E-W Highway
Silver Springs, MD 20910

Phone: (301) 713-3115

Southeastern Regional Climate Center
1201 Main Street, Suite 1100
Columbia, SC 29201
Phone: (803) 737-0849

Insurance

In coastal areas, special building requirements must often be met to obtain wind-storm insurance or affordable flood insurance. To find out the requirements for your area, check with your insurance agent. Further information is available from:
Flood Insurance Administration
500 C Street, SW
Washington, DC 20472

Director
Natural and Technological Hazards Division
FEMA Region IV Office
Suite 700
1371 Peachtree Street, NE
Atlanta, GA 30309-3108
Phone: (404) 853-4400

South Carolina Land Resources and Conservation Districts Division
2221 Devine Street, Suite 222
Columbia, SC 29205
Phone: (803) 734-9103

For V-zone coverage or a request for an individual structure rating, contact:

National Flood Insurance Program
Attn: V-Zone Underwriting Specialist
P.O. Box 6468
Rockville, MD 20849-6468
Phone: (800) 638-6620

For flood map requests and forms, contact:

FEMA Flood Map Distribution Center
6930 (A-F) San Thomas Road
Baltimore, MD 21227-6627
Phone: (800) 358-9616

Your insurance agent or community building
inspector should be able to show you the location of your building site on the Flood Insurance Rate Map (FIRM) and help you determine
the elevation required for the first floor to be
above the 100-year flood level. If they cannot
provide this information, request the FIRM for
your area from FEMA at the above address.
Note that a flood policy under the National
Flood Insurance Program is separate from
your regular homeowner's policy.

The Insurance Institute for Property Loss
Reduction is an umbrella group of private insurers that offers many different types of services for member companies. The institute has
an extensive publications list, including information on property insurance research, publications, videos, and computer programs. Contact:

Insurance Institute for Property Loss
Reduction

73 Tremont Street, Suite 510
Boston MA, 02108-3910
Phone: (617) 722-0200

Land Acquisition

Anyone acquiring property or a condominium,
whether it is in a subdivision or not, should
consider the following points. (1) Prospective
buyers of property next to dredged canals
should make sure that the canals are designed
for adequate flushing to keep them from becoming stagnant. Requests for federal and
state permits to connect extensive canal systems to navigable water are frequently denied.
(2) Description and survey of land in coastal
areas is very complicated. Old titles granting
fee-simple rights to property below the high-
tide line may not be upheld in court; titles
should be reviewed by a competent attorney
before they are transferred. A boundary described as the high-water mark may be impossible to determine. (3) Ask about the provision
of sewage disposal and utilities, including water, electricity, gas, and telephone. (4) Be sure
any promises of future improvements, access,
utilities, additions, common property rights,
etc., are in writing. (5) Be sure to visit the
property and inspect it carefully before buying
it. (See *Planning and Land Use* and *Subdivisions,* below.)

Maps

Maps are useful to planners and managers and
may be of interest to individual property owners. Topographic, geologic, and land-use maps
and orthophoto quadrangles are available
from the following:

Earth Science Information Center
U.S. Geological Survey
12201 Sunrise Valley Drive
Reston, VA 22092
Phone: (703) 860-6045

Request a free index to the type of map desired
(e.g., "Index to Topographic Maps of South
Carolina") and use it to order specific maps.
Similar maps are available from:

State Geologist
South Carolina Geological Survey
5 Geology Road
Columbia, SC 29210
Phone: (803) 896-7708

For evacuation maps, call your county Department of Emergency Preparedness (see *Hurricane Information*). For flood zone maps, see
Insurance. For planning maps, call or write
your local county commission. For soil maps
and septic suitability, see *Soils.*

Nautical charts in several scales contain navigation information on South Carolina's coastal

waters. A nautical chart index map is available from:

Distribution Branch N/CG33
National Oceanic Service
National Oceanic and Atmospheric Administration
Riverdale, MD 20737-1199
Phone: (301) 436-6990

For county road maps, contact:

South Carolina State Department of Highways and Public Transportation
P.O. Box 1498
Columbia, SC 29202
Phone: General information (803) 737-1130
Map sales (803) 737-1501

Marine and Coastal Zone Information

In addition to the government agencies listed by topic in this appendix, many private agencies, laboratories, and educational institutions are sources of valuable reports and descriptive pamphlets and can answer your questions; many also sponsor coastal programs, including lectures, seminars, and film series. The following list is not meant to be complete, only to suggest the range of possibilities:

South Carolina Sea Grant Consortium
287 Meeting Street
Charleston, SC 29401
Phone: (803) 727-2078

National Sea Grant Depository
Pell Library Building
University of Rhode Island
Narragansett Bay Campus
Narragansett, RI 02882-1197
Phone: (401) 792-6114
e-mail: cmurray@gsosunl.gso.uri.edu

The depository was established in 1970 to serve as an archive of all Sea Grant–funded documents. It also includes library services.

Coast Alliance
218 D Street, SE
Washington, DC 20003
Phone: (202) 546-9554

University of South Carolina
Belle W. Baruch Institute for Marine Biology and Coastal Research Columbia, SC 29208
Phone: (803) 777-5288

Grice Marine Laboratory
College of Charleston
205 Ft. Johnson
Charleston, SC 29412
Phone: (803) 763-5550

The South Carolina Wildlife and Marine Resources Department publishes the bimonthly magazine *South Carolina Wildlife,* which includes articles about the coast and sometimes devotes an entire issue to it (e.g., vol. 27, no. 4, July–August 1980). For subscription information, write:

South Carolina Department of Natural Resources
P.O. Box 167
Columbia, SC 29202

The National Audubon Society's bimonthly *Audubon Magazine* often includes articles that deal with coastal problems as well as wetland protection and coastal flora and fauna. For subscription information, write:

National Audubon Society
Southeast Regional Office
P.O. Box 1268
Charleston, SC 29402
Phone: (803) 577-7100

Also check with the Clemson University Extension Office in the coastal county in question as well as the offices of the appropriate city or county planning commission.

Movies and Audiovisual Materials

For general information, contact the South Carolina Sea Grant Program. In the past, audiovisual materials on coastal topics have been available through the following:

Instructional Service Center
University of South Carolina
700 Assembly Street, Sublevel Floor 1
Law Center
Columbia, SC 29208
Phone: (803) 777-6430

Slide-tape programs and films on coastal topics may be available from:

National Audubon Society
Southeast Regional Office
P.O. Box 1268
Charleston, SC 29402
Phone: (803) 577-7100

Films on North Carolina's barrier islands (*Waterbound, An Act to Protect,* and *The Currituck Film*) can be borrowed or bought from:

UNC Sea Grant Program
Box 8605
North Carolina State University
Raleigh, NC 27695-8605
Phone: (919) 515-2454

The Atlantic's Last Frontier examines Virginia's barrier island coast; information on this film is available from:

The Nature Conservancy
1815 N Lynn Street
Arlington, VA 22209
Phone: (703) 841-5300

The Beaches Are Moving is a 60-minute videotape hosted by Dr. Orrin Pilkey of Duke University. Produced in 1990 for public television station WUNC-TV in Chapel Hill, the video examines the conflicts of coastal development and explains what we can do to have our

beach and save it too. As of this writing, a new video, *Living on the Edge,* was nearing its final production stages. To order the videocassettes, contact:

Environmental Media
P.O. Box 1016
Chapel Hill NC 27514
(919) 933-3003

Parks and Recreation

Coastal state parks include Myrtle Beach, Huntington Beach, Charles Towne Landing, Edisto Beach, and Hunting Island. For general information on parks, contact the following:

Division of Parks
Department of Parks, Recreation, and Tourism
1205 Pendleton Street
Columbia, SC 29201
Phone: (803) 734-0122

The Cape Romain National Wildlife Refuge islands are accessible only by boat. Public boat ramps are located at Moore's Landing, Buck Hall, and McClellanville. Do not enter areas marked "Area Closed."

Cape Romain National Wildlife Refuge
5801 Highway 17 N
Awendaw, SC 29429
Phone: (803) 928-3368

The number of visitors to the Tom Yawkey

Wildlife Center (South Island, Cat Island, and North Island) is limited. Requests should be made 30 days in advance to:

Project Leader
Tom Yawkey Wildlife Center
Route 2, Box 181
Georgetown, SC 29440
Phone: (803) 546-6814

Addresses of the chambers of commerce for coastal cities as well as various private resorts are available from the South Carolina Department of Parks, Recreation, and Tourism (given above).

Historic Fort Sumter in Charleston Harbor and Fort Moultrie on Sullivans Island are national monuments under the administration of the National Park Service and are open to the public daily.

Planning and Land Use (see also *Coastal Zone Management*)

South Carolina Department of Health and Environment Control
Office of Ocean and Coastal Resource Management
4130 Faber Place, Suite 300
Charleston, SC 29405
Phone: (803) 744-5838

South Carolina Department of Commerce
Division of State Development

P.O. Box 927
Columbia, SC 29202
Phone: (803) 737-0400

For specific information on your area, check with your local town or county commission. Most local governments have planning boards that answer to the commission and have copies of existing or proposed land-use plans available for examination.

Roads and Property Access

The South Carolina Department of Highways and Public Transportation is not required to furnish access to all property owners. Before buying a piece of land, determine whether access rights and roads will be provided. If you are connecting a driveway to a state-maintained right-of-way, you will probably need a permit from the highway department. Contact a state or county road official:

South Carolina Department of Transportation
P.O. Box 1498
Columbia, SC 29202
Phone: (803) 737-1130

Sanitation and Septic System Permits (see also *Water Resources*)

If a property has no access to a sewer system, it is usually necessary to obtain a permit for a septic system from the local health department before a construction permit can be issued.

Such a permit will be issued only if the soil is suitable for a septic system. Old marsh muds and peaty soils are usually unsuitable. Likewise, if your property does not have access to a municipal water system, you will need a well. Check with the county health department to determine the quality of the local groundwater. Make sure that the design and location of your septic system will safeguard your water supply.

Activity resulting in effluent discharge or runoff into surface waters requires certification from the state water pollution control agency that the proposed activity will not violate water quality standards. For information, contact:

South Carolina Department of Health and Environmental Control 2600 Bull Street
Columbia, SC 29201
Phone: (803) 734-5000

A permit for a sewage disposal structure or any other structure in navigable waters must be obtained from the U.S. Army Corps of Engineers and the DHEC. More information is available from the following:

Permits Division
U.S. Army Corps of Engineers
P.O. Box 919
Charleston, SC 29402-0919
Phone: (803) 727-4330

A permit from the U.S. Environmental Protection Agency is required for any discharge

into navigable waters. Recent judicial interpretation of the Federal Water Pollution Control Amendments of 1972 extends the federal jurisdiction for protection of wetland above the mean high-water mark. Federal permits may now be required to develop land that is occasionally flooded by water draining indirectly into a navigable waterway. Information may be obtained from:

Environmental Protection Agency
Region IV
345 Courtland Street, NE
Atlanta, GA 30365
Phone: (404) 347-3004

Soils (see also *Sanitation and Vegetation*)

Soil type is important in terms of (1) the type of vegetation it can support, (2) the type of construction it can withstand (e.g., loading, support of piling), (3) its drainage characteristics, and (4) its ability to accommodate septic systems. The following agencies cooperate to produce a variety of maps and reports useful to property owners:

Natural Resources Conservation Service
1835 Assembly Street, Room 950
Columbia, SC 29201
Phone: (803) 765-5681

Clemson University Extension Office: see the white pages of your telephone directory for the local number.

Soil and Water Conservation District Office: see the white pages of your telephone directory for the local number.

Your community or county health department usually can provide soils information relative to construction and septic permits or can refer you to another agency for specific soil information.

Subdivisions

Subdivisions containing more than 100 lots and offered in interstate commerce must be registered with the Office of Interstate Land Sales Registration (as specified by the Interstate Land Sales Full Disclosure Act). Prospective buyers must be provided with a property report. This office also produces a booklet titled *Get the Facts Before Buying Land* for people who wish to invest in land. Information on subdivision property and land investment is available from:

Office of Interstate Land Sales Registration
U.S. Department of Housing and Urban Development
451 7th Street, SW, Room 9160
Washington, DC 20410
Phone: (202) 708-0502

Vegetation

Information on vegetation may be obtained from your local Soil and Water Conservation district office. For information on the use of grass and other plantings for stabilization or aesthetics, consult the publications listed in appendix D under "Vegetation." *Seacoast Plants of the Carolinas for Conservation and Beautification* (ref. 114) is particularly useful.

Water Resources

The South Carolina Department of Health and Environmental Control is responsible for regulation, permit issuance, construction information, and studies on water availability, quality, pollution control, well development, waste disposal, and similar problems. Contact the appropriate division or bureau for particular problems or information. With the exception of regional offices in some coastal counties, all the offices are at the same address. Division responsibilities are listed below with individual telephone numbers:

South Carolina Department of Health and Environmental Control 2600 Bull Street
Columbia, SC 29201
Phone: (803) 734-5000

Division of Industrial Wastewater (regulation and permits for ponds, lagoons, and land disposal of effluents)
Phone: (803) 734-5253

Division of Domestic Wastewater (regulation of septic systems)
Phone: (803) 734-5300

Division of Drinking Water (regulation and permits for public water-supply wells)
Phone: (803) 734-5310

Division of Solid and Hazardous Waste Management (construction, regulation, and permitting for solid waste disposal sites)
Phone: (803) 896-4000

The OCRM has final permitting and certification authority in its zone of coastal jurisdiction. Barrier island and nearshore mainland development should always be checked through the OCRM's permit system:

South Carolina Department of Health and Environmental Control
Office of Coastal Resource Management
4130 Faber Place, Suite 300
Charleston, SC 29405
Phone: (803) 744-5838

Other agencies that provide information on both surface and groundwater availability and quality include the following:

U.S. Geological Survey, Water Resources Division
720 Gracern Road
Stephenson Center, Suite 129
Columbia, SC 29201-7651
Phone: (803) 750-6100

State Geologist
South Carolina Geological Survey
5 Geology Road

Columbia, SC 29210
Phone: (803) 896-7708

Weather

General and current information may be obtained most easily from local radio, television, and newspapers.

Wildlife

National Audubon Society
Southeast Regional Office
P.O. Box 1268
Charleston, SC 29402
Phone: (803) 577-7100

South Carolina Wildlife and Marine Resources
Department
P.O. Box 12559
Charleston, SC 29412
Phone: (803) 762-5000

Zoning (see *Building Codes and Zoning*)

APPENDIX C   The Beachfront Management Act, an Amendment to the Coastal Tidelands and Wetlands Act

The following is a summary of the Beachfront Management Act that explains the law in terms familiar to the layperson. It is not intended to to be used for legal purposes. The reader should refer to the original document for exact wording or clarification of the law. The Office of Coastal Resource Management (OCRM) will provide a copy of the Coastal Zone Management Act, to which the Beachfront Management Act is an amendment. The reader is urged to seek legal counsel when necessary.

Section 48-39-250: Legislative Declaration of Findings Concerning the Beach-Dune System

Beginning with "The General Assembly finds that:" this section discloses that the beach-dune system of South Carolina is extremely important and serves the following functions: it protects life and property and provides a basis for the tourism industry, habitat for numerous species, and a healthy environment.

Vegetation of the beach-dune system is unique and extremely important to the vitality and preservation of the system.

Many miles of this unique system are critically eroding; however, prior to 1988, chapter 39 of Title 48, "Coastal Tidelands and Wetlands," did not provide adequate jurisdiction to enable the South Carolina Coastal Council to protect the integrity of the beach-dune system. Development has jeopardized the system, accelerated erosion, and endangered property.

Shoreline armoring has not proven effective and has been shown to accelerate erosion.

Erosion and accretion are natural processes which require space. This space can be provided only if building too close to the beach-dune system is discouraged and retreat from it is encouraged.

Inlet and harbor management practice may deprive downdrift beach-dune systems of their natural sand supply.

It is in the state's interest to protect South Carolina's beaches and promote beach access, but present funding for this is inadequate.

There is no coordinated state policy for poststorm emergency management of the beach-dune system.

A long-range comprehensive beach management plan is needed to minimize man's adverse impact on the system.

Section 48-39-260: Legislative Declarations of State Policy Concerning the Beach-Dune System

In recognition of its stewardship responsibilities, the policy of South Carolina is to

1. Protect, preserve, restore and enhance the beach-dune system
2. Create a comprehensive, long-range beach management plan and require the same of local governments
3. Severely restrict the use of shoreline armor
4. Promote the use of "soft" technologies
5. Promote beach nourishment where economically feasible
6. Preserve and enhance public access
7. Involve local governments in long-range planning and management
8. Establish procedures and guidelines for emergency storm response

Section 48-39-270: Beach-Dune Definitions

Section 48-39-270 provides definitions of the terms mentioned in the act, including *erosion control devices, habitable structures, beach nourishment, active beach,* and others. The beach-dune system includes all land from the mean high-water mark of the Atlantic Ocean landward to the setback line described in section 48-39-280. The two types of regulated shoreline are defined as standard erosion zones and inlet erosion zones.

Section 48-39-280: Procedure for the Establishment of Baselines and Setback Lines

Section 48-39-280 (A) establishes the basis for a 40-year retreat policy, defining the baselines and setback lines and giving a schedule for their reestablishment, which must take place not less than every 8 years and not more than every 10 years. In order to establish baselines and setback lines, the OCRM must install a network of survey monument stations at 2,000-foot intervals or less along the developed coast.

Note: This survey network was established

and has been in use since 1986.

The three baseline definitions are explained as follows:

1. The baseline for a standard erosion zone, a relatively straight shoreline unaffected by inlets, is the crest of the primary dune. In the absence of a primary dune, the council shall calculate where that dune would ideally be.

2. The baseline for an unstabilized inlet erosion zone—i.e., one not stabilized by jetties, groins, or other structures—shall be the most landward position of that shoreline in the last 40 years, unless the best available scientific and historical data indicate the shoreline is unlikely to return to this position.

3. The baseline for a stabilized inlet erosion zone—i.e., a shoreline stabilized by jetties, terminal groins, or other structures—shall be the crest of the primary dune in that zone.

This section also describes when and how a baseline may be moved seaward in response to a successful beach nourishment program.

Finally, section 48-39-280 (A) provides a means for a formal review if a property owner believes there is an error in the official baseline, setback line, or erosion rate.

Section 48-39-280 (B) establishes that the retreat policy in part A is implemented by the establishment of a setback line which, measured from the baseline, is 40 times the annual erosion rate, or not less than 20 feet. This ensures that a highly erosional beach will have an appropriately wide setback area. Conversely, a nonerosional beach will have a minimal setback. For example, a beach with 2 feet per year erosion will have a setback line 80 feet from the dune crest. If the annual erosion rate is 5 feet, the setback area is 40 times 5 feet, or 200 feet wide. (Remember that jurisdiction lines and erosion rates are periodically revised.)

Section 48-39-290: Alterations Seaward of Baseline and Setback line

Part A of this very long and complex section of the Beachfront Management Act describes what may and may not be constructed, reconstructed, or altered within the state's permitting jurisdiction. It first provides an exemption for the following seven kinds of construction, although only the first may be done without a permit:

1. Wooden walkways for beach access, less than 6 feet wide
2. Small wooden docks
3. Fishing piers
4. Golf courses
5. Landscaping
6. Specific structures detailed in subsection D
7. Pools landward of an erosion control structure

Habitable Structures

In some cases, a new house may be built seaward of the setback line if there is insufficient room on the lot in question landward of the setback line. Certain conditions and restrictions apply, however, and compliance with these must be certified in writing to the OCRM. For example, no erosion control structure may be incorporated as part of the building and no primary dune may be affected, and houses are restricted to 5,000 square feet of heated space.

Previously existing houses may be maintained and repaired. They may also be enlarged, but only subject to the same restrictions that apply to new houses. Buildings destroyed by natural forces may be rebuilt, but there may be conditions or restrictions. Only OCRM staff members or attorneys are qualified to answer specific questions regarding structures in the setback area.

Erosion Control Structures

No new erosion control structures may be built seaward of the setback line except to protect a public road which existed on the date this act became effective. Existing erosion control structures may be maintained but not enlarged, strengthened, or rebuilt. If such a structure is destroyed, it must be removed from the beach by the owner.

Whether or not a seawall is destroyed must be determined by a registered professional engineer working for the ocrm. If the owner disagrees with that determination, he may obtain an assessment by another registered engineer. If the two assessments differ, the engineers must select a third engineer to perform the assessment. If they cannot agree, then the local county clerk of court must select an engineer, and that engineer's damage assessment is conclusive.

A detailed formula is given in this subsection as to the method by which the engineer's damage assessment is made. Using the engineer's figures, a structure is determined to be destroyed if the damage to it exceeds 80 percent, through June 30, 1995. From then until June 30, 2005, the damage must exceed two-thirds. Thereafter, any erosion control structure damaged more than 50 percent is considered destroyed.

Pools

No new pool may be built seaward of the setback line unless the site is landward of an erosion control structure. Normal maintenance and repairs are allowed for pools.

Pools that were destroyed may be rebuilt no larger and must be moved landward if possible. Pools may not be reinforced to function as seawalls. A new pool may be built to replace an old, undamaged one, but the new pool can be no larger or stronger than the old one.

Any and all other construction between the baseline and the setback line requires a permit.

However, the council may issue general permits for activities which advance the goals of this law (for example, sand fencing or dune vegetation) as indicated in sections 250 and 260.

Work begun by the effective date of this act may continue. This law does not prohibit permitted fishing piers or beach access walkways.

Special Permits

If an applicant requests a permit to build or rebuild a structure other than an erosion control structure, the OCRM may grant this permit, but the following conditions apply:

1. The site is not or does not include a primary dune.
2. The site is not an active beach.
3. If the site erodes and becomes active beach, the owner must remove the structure in question.

In making its decision the OCRM must determine that this permitted activity is not detrimental to the health, safety, and welfare of the public. Reasonable additional safeguards and conditions may be applied in order to advance the purposes of the act.

The DHEC's decisions regarding special permits are legal matters, and anyone wishing to appeal a decision should obtain professional legal counsel. Appeals to the OCRM's decisions on special permits are the province of the circuit court only. There, professional legal counsel is required.

(E) Section III Exemption

This subsection exempts areas where beach erosion is attributed to a federal navigation project from all jurisdictional definitions of section 280 and all previous stipulations and restrictions of section 290. This link must be established by a study under section III of the Federal Rivers and Harbors Act of 1968, as amended by the Water Resources Development Act of 1986, and approved by the U.S. Army Corps of Engineers. The only place, to date, where this situation applies (and probably could apply) is Folly Beach.

Because of this exemption there are no setback lines on Folly Beach. The baseline is the line of erosion control devices. The state's jurisdiction begins seaward of this "seawall line." There are a few situations where critical beach areas exist landward of the baseline, and the city of Folly Beach is obligated by the Beachfront Management Act to protect these areas.

It is important to note that Folly Beach is exempt only from the conditions set forth by sections 280 and 290. The exemption thus essentially negates the "retreat" policy which is the intent of the Beachfront Management Act. However, the exemption requires that the policy of retreat be addressed in Folly's Local Shorefront Management Plan. There are completely adequate controls in other sections of the act to protect Folly's beach from development and to promote activities that enhance beach quality and beach access.

Section 48-39-300, Section III Study Exemption: Erosion Control Devices

Erosion control devices on Folly Beach may not be made larger, moved seaward, or rebuilt out of materials different from that of the original structure.

Section 48-39-305: Judicial Review of Takings Claims

Property owners affected by the law may petition the circuit court to determine if the use of their land is so restricted that it is an unreasonable exercise of police power and constitutes a taking without compensation. If the court finds in favor of the property owner, then it must require the state either to issue the necessary permits, to order that the prohibition does not apply to that property, or to provide reasonable compensation or the payment of costs and reasonable attorney's fees, or both. Either party may appeal this decision.

Section 8-39-310: Destruction of Beach-Dune Vegetation Is Prohibited

Destruction of beach or dune vegetation is prohibited seaward of the setback line unless, for some permitted activity, there is no feasible alternative. When destruction of vegetation is unavoidable, new vegetation must be planted, where possible, to mitigate the damage.

Section 48-39-320 (A): State Beachfront Management Plan

The Coastal Council [now the OCRM] is responsible for the creation of a long-range and comprehensive beach management plan for South Carolina's Atlantic shoreline. This plan shall include the following:

1. Development of a database necessary to make informed and scientifically based decisions concerning the maintenance or enhancement of the beach-dune system.
2. Development of guidelines and their coordination with appropriate agencies and local governments to accomplish the following:

   (a) beach-dune restoration and nourishment
   (b) development and enhancement of public beach access
   (c) maintenance of a stable dry-sand beach
   (d) protection of dunes seaward of the setback line
   (e) protection of endangered and threatened species and habitats
   (f) regulation or prohibition of vehicular traffic on beaches
   (g) a mitigation policy for permitted construction impacts on vegetation, public access, sand resources, etc.
3. Recommendations for funding programs to achieve the goals of the state's beach management plan.
4. Development of a program of public education regarding the beach-dune system.

This program is to be developed in coordination with the South Carolina Educational Television Network and the Department of Parks, Recreation, and Tourism.

5. Assistance to local governments to develop beach management plans.

Section 48-39-330: Real Estate Disclosure Statement

A contract for the sale or transfer of property affected by the setback line must contain a deed disclosure statement. This statement must delineate the relationship of the property and any habitable structures to the baseline and setback line, and be referenced to the State Plane Coordinate System. The disclosure statement must contain the current erosion rate for that location.

Section 48-39-340: Funding for Local Governments to Provide Beachfront Management

This section briefly describes how funding for beachfront management shall be distributed fairly. Consideration must go to the size of the beach, its need for management, the costs and benefits of expenditures, and the best interest of the beach-dune system.

Section 48-39-350: Local Beachfront Management Plans

Subsections A and B are described below.

A. Each local government is required to prepare a local comprehensive beach management plan, which must be approved by the council.

This plan must contain the following:

1. An inventory of beach profile data and historical erosion rate data for all local beach areas. This data is provided by the OCRM.
2. An inventory of public beach access and parking and a plan for improving these.
3. An inventory of all structures in the setback area.
4. An inventory of turtle nesting and important habitats of the beach-dune system and a protection and restoration plan if necessary.
5. A conventional zoning and land-use plan consistent with the purposes of this chapter.
6. An analysis of beach erosion control alternatives.
7. A drainage plan for the beach area.
8. A postdisaster plan for cleanup, maintaining essential services, protecting public health, emergency building ordinances, and the establishment of priorities consistent with this chapter.
9. A detailed strategy for achieving the goals of this chapter; consideration must be given to relocating buildings, removing erosion control structures, and relocating utilities.
10. A detailed plan to preserve and enhance public access. This plan must be updated at least every five years.

B. If a local government fails to implement a beachfront management plan by July 1, 1992, or to enforce it, the OCRM must impose and implement the State Comprehensive Beach Management Plan instead. In such a case, the local government automatically loses its eligibility to receive state-generated or shared funds for any beach-dune system projects.

Section 48-39-355: Documentation for Activities Allowed by This Chapter

No permit is required for an activity specifically authorized in this chapter. However, the OCRM may require documentation to establish that the activity is in compliance with this law.

Section 48-39-360: Limit on Jurisdiction for Sections 280–360

The landward limit of the state's beachfront jurisdiction is one-half mile from the mouth of an inlet. The mouth of an inlet is defined as the narrowest part of the inlet.

Draft Language for Real Estate Disclosure Statements

1. This property is subject to regulation of use by the Coastal Management Act, section 48-39-10, et seq., 1976 S.C. Code of Laws, as amended, and part of (all of) this property is seaward of the setback line/and minimum setback line/and baseline/and has an erosion rate of _____ feet per year, all as adopted by the South Carolina Department of Health and Environmental Control's Office of Ocean and Coastal Resource Management. This property (Part of this property) is also within the velocity zone as determined by FEMA.

2. More specifically, the setback line is _____ feet (from _____ feet to _____ feet) from the seaward property line; the baseline is _____ feet (from _____ feet to _____ feet) from the seaward property line. The velocity zone is _____ feet wide (from _____ feet to _____ feet wide) starting at the seaward property line and moving landward. The seaward corners of the habitable structures on this property are located _____ feet, _____ feet, etc., from the seaward property line.

3. This information is shown with more particularity on that certain plat made by _____, dated _____, filed in Plat Book _____, page _____, Clerk of Court's Office for _____ County, a copy of which is attached hereto/reference to which is hereby prayed for a more complete disclosure.

Most people will need to have the plat analyzed by a qualified individual, and that information (and possibly more) will be needed to determine whether the use of all of the terms in the paragraphs is necessary. Whether or not the baseline or minimum setback line is on the property along with the setback line is a case-by-case determination.

APPENDIX D   Useful References

The publications cited below are listed by subject and arranged in the approximate order of their appearance in the book. A brief annotation accompanies most references. Many of the publications in this appendix are available either free or at a nominal cost. We encourage the reader to take advantage of them.

Some of the publications will be easier to find than others. We include the publisher's address with some entries; the reference desk at your library can help you find others. University libraries are the best place to find scientific literature (books and journal articles). If you are lucky enough to live near a major university, you should have no trouble obtaining almost everything listed here. Even smaller university libraries should be able to retrieve articles and books for you via interlibrary loan. Popular books should be available at general bookstores; if not, any local bookstore should be able to order them for you. Government publications are available from the agency or from the U.S. Government Printing Office in Washington, D.C. See appendix B for addresses of all the agencies, government and private, involved with the aspects of the coastal zone covered in this book. New sources of information are published every day. Contact the agencies and publishers listed in them for new materials.

Another source of information is the "information superhighway." More and more agencies and organizations are creating their own home pages on the World Wide Web. If you have access to a computer with internet hookup, a quick search with a network search program should provide you with loads of information. Document search-and-retrieval capabilities are available on the internet as well. Your local computer store should be able to point you in the right direction, as should libraries.

To save space, we use agency abbreviations as much as possible in the following list. The abbreviations we have used are given below:

ASBPA	American Shore and Beach Preservation Association
BAREPP	Bay Area Regional Earthquake Preparedness Project
CERC	Coastal Engineering Research Center
CERF	Coastal Education and Research Foundation
CSO	Coastal States Organization
DHEC	South Carolina Department of Health and Environmental Control
DOE	U.S. Department of Energy
EPA	U.S. Environmental Protection Agency
FEMA	Federal Emergency Management Agency
FIA	Federal Insurance Administration
FSBPA	Florida Shore and Beach Preservation Association
NAS	National Academy of Sciences
NFIP	National Flood Insurance Program
NOAA	National Oceanic and Atmospheric Administration
NOAA-OCRM	Federal Office of Ocean and Coastal Resource Management, under NOAA
NRC	National Research Council
OCRM	South Carolina Office of Ocean and Coastal Resource Management, under the DHEC (referred to as DHEC-OCRM); formerly the South Carolina Coastal Council
USACOE	U.S. Army Corps of Engineers
USGS	U.S. Geological Survey

History

1. *The Grand Strand,* by Nancy Rhyne, 1981. Subtitled *An Uncommon Guide to Myrtle Beach and Its Surroundings,* this short book provides entertaining reading for persons living and vacationing on the Horry County coast. Brief histories of the various beach communities offer some insight on coastal development and the impact of past storms. Published by East Woods Press, 820 East Boulevard, Charlotte, NC, 28203.

2. *Pawleys—As It Was,* by Celina Vaughan, 1975, is a short history of the development of a single barrier island from plantation days to the present. It includes a chapter on hurricanes. Privately printed by the Hammock Shop, Pawleys Island, SC.

3. *Pawleys Island . . . a Living Legend,* by C. K. Prevost and E. L. Wilder, 1972, is an interesting narrative on the history of an old resort, including accounts of nineteenth-century hurricanes on the island. Printed by the State Printing Company, Columbia, SC.

4. *The History of Georgetown County, South Carolina,* by G. C. Rogers, Jr., 1970. This 565-page book includes an outline of the Proprietor system as well as historic insight into the development of the Santee Delta islands, North Island, and Debordieu Island, and subsequent storm destruction. Published by the University of South Carolina Press, Columbia 29208.

5. *Early American Hurricanes, 1492–1870,* by D. M. Ludlum, 1963. This excellent summary of the stormy history of the Atlantic and Gulf coasts provides a lesson on the frequency, intensity, and destructive potential of hurricanes. Published by the American Meteorological Society, Boston, MA.

6. *America's Lighthouses, an Illustrated History,* by Francis Ross Holland, Jr., 1972, includes photographs, drawings, and histories of hundreds of lighthouses in the United States and Europe, as well as the developmental history of federal lighthouse administration. Published by Dover Publications, New York, NY.

7. *South Carolina,* by Henry Leifermann, 1995. More than a travel guide, this well-written handbook (321 pp.) provides a sense of South Carolina's history and socioeconomic setting. Of particular interest are the sections on the Grand Strand and the sea islands, but the reader may be equally drawn to the coastal plain and piedmont. Some good insights into the politics of coastal communities emerge. Published by Compass American Guides, Oakland, CA, and available in bookstores.

8. *To Take Charleston: The Civil War on Folly Island,* by James William Hagey, 1993. This is an excellent account of the Union Army's occupation of Folly Island during the Civil War siege of Charleston. This highly illustrated work, drawn primarily from regimental archives, culminates in the battle for Fort Wagner and the fall of Fort Sumter and Charleston. A fascinating read. Published by Pictorial Histories Publishing Company; available in libraries and bookstores.

Hurricanes and Storms

Hurricanes, General

9. *South Carolina Hurricanes,* by J. C. Purvis and H. Landers, 1973, and its supplement, *South Carolina Hurricanes, 1950–1979,* by J. C. Purvis, 1980. A descriptive listing of tropical cyclones that have affected South Carolina since the earliest recorded history of the region. Provides striking but nontechnical data indicating that destructive hurricanes are not rare events in the state. Published by the South Carolina Emergency Preparedness Division, Office of the Adjutant General, Columbia, SC 29201.

10. *Investigation on Hurricanes and Associated Problems along the South Carolina Coast,* by USACOE, 1957. This appraisal report summarizes the history of South Carolina hurricanes and discusses the destructive impacts of the August 11, 1940; October 15, 1954 (Hazel); August 11, 1955 (Connie); and August 17, 1955 (Diane) storms. Available for inspection at the USACOE's Charleston District Office library.

11. *Hurricanes and Coastal Storms,* edited by Earl Baker, 1980. These technical papers dealing with hurricane and storm awareness, evacuation, and property damage mitigation were presented at a 1979 national conference in Orlando, Florida. Good reading for planners, developers, and coastal community officials. Available from the Marine Advisory Program, University of Florida, Gainesville 32611.

12. "The Hurricane Disaster Potential Scale," by R. H. Simpson, published in 1974 in the magazine *Weatherwise* (vol. 27, p. 169), is the first description and presentation of what is now commonly known as the Saffir/Simpson scale. *Weatherwise* is a popular meteorology magazine that contains a lot of good scientific information, yearly summaries of weather, and fun facts about weather and the history of weather forecasting and observation.

13. *Atlantic Hurricanes,* by G. E. Dunn and B. I. Miller, 1960, discusses at length hurricanes and associated phenomena such as storm surge, wind, and sea action. Includes a detailed account of Hurricane Hazel (1954) and sug-

gestions for pre-and posthurricane procedures. An appendix includes a list of hurricanes that have affected the Carolinas. Published by Louisiana State University Press, Baton Rouge 70803.

14. *The Hurricane and Its Impact,* by R. H. Simpson and H. Riehl, 1981, is a classic, often-cited text that discusses hurricane origins; the impact of wind, waves, and tides; assessment and risk reduction; awareness and preparedness; and prediction and warnings; includes informative appendixes. Published by Louisiana State University Press, Baton Rouge 70893.

15. *The Deadliest, Costliest, and Most Intense United States Hurricanes of This Century (and Other Frequently Requested Hurricane Facts),* by Paul J. Hebert and Glenn Taylor, 1988, is NOAA Technical Memorandum NWS-NHC-18. Now in its fifth printing, this pamphlet contains hurricane facts and several tables summarizing deaths, costs, and hurricane intensities. In addition, tables portray hurricane activity by year and the most recent major hurricanes to affect each state. Available from NOAA, National Weather Service.

16. *"Hurricane!" A Familiarization Booklet,* by NOAA, 1993, is a descriptive and nontechnical overview of U.S. hurricanes. Includes sections on hurricane anatomy, storm surge, forecasting, lists of the most intense and destructive hurricanes through 1992, and a hurricane checklist. The 36-page document, NOAA PA 91001, is available through NOAA.

17. *Storm Surge and Hurricane Safety with*

North Atlantic Tracking Chart, by NOAA, 1982. This brochure has brief, simple discussions of the components of storm surge and hurricanes. Outlines method of tracking hurricanes and provides a detailed checklist. It is document 1982 0-379-175, available from the U.S. Government Printing Office.

18. "The Forgotten Hurricane," by Jeff Rosenfeld, summarizes the impact of the August 27–28, 1893, hurricane on South Carolina's sea islands and the Savannah, Georgia, area. Published in *Weatherwise* (August–September 1993), pp. 13–18.

19. *Hurricane Evacuation Study for South Carolina,* G21-*Atlas of Maximum Envelope of High Water for Hurricanes Affecting the South Carolina Coast.* Part 2: *Hurricanes Moving from All Directions (MOMS),* by J. C. Purvis and G. Yarbrough, 1986. This 53-page atlas of maps of coastal South Carolina depicting maximum potential inundation by storm surges from hurricanes of various categories was produced for use by emergency personnel in the evacuation decision process. Available from the South Carolina Emergency Preparedness Division, Columbia.

20. *The Weather Book,* by Jack Williams, 1992, explains weather phenomena in easily understood language, using many color diagrams and photographs. Among the topics covered are hurricanes, thunderstorms, frontal storms such as northeasters, floods, precipitation, the greenhouse effect, and the ozone layer. Published by Vintage Books, Random

House, New York, NY.

21. *Against the Wind* is a six-page brochure, published in 1993, that contains summary information about protecting your home from hurricane wind damage. It was developed jointly by the American Red Cross, FEMA, Home Depot, the National Association of Home Builders of the United States, and the Georgia Emergency Management Agency. It briefly discusses roof systems, exterior doors and windows, garage doors, and shutters. Available from the American Red Cross (publication ARC-5023) and FEMA (FEMA-247).

Hurricane Hugo

22. *Storm-Tide Elevations Produced by Hurricane Hugo along the South Carolina Coast, September 21–22, 1989,* by R. Erik Schuck-Kolben, 1990. USGS Open-File Report OF 90-0386, prepared in cooperation with FEMA, 31 sheets.

23. The professional journal *South Carolina Geology* devoted an entire issue to Hurricane Hugo one year after the storm. Volume 33, no. 2, contains five articles by nine authors covering shoreline change on Sullivans Island, Isle of Palms, and the Grand Strand; sediment budget changes for Isle of Palms; impacts on the Charleston shoal area; and general effects on shoreline landforms. This journal is a publication of the South Carolina Geological Survey.

24. *Shore and Beach,* vol. 58, no. 4 (October 1990), published by the ASBPA, is devoted to

scientific articles about Hurricane Hugo.

25. *Impacts of Hurricane Hugo: September 10–22, 1989,* edited by Charles W. Finkl and Orrin H. Pilkey, 1991, was published by the CERF as *Journal of Coastal Research,* Special Issue 8. Its 356 pages contain 24 scientific papers describing various aspects of Hugo's meteorology and effects on the Virgin Islands, Puerto Rico, and South Carolina. Most of the papers deal with South Carolina. Available through university libraries or from CERF, 4310 NE 25th Avenue, Fort Lauderdale, FL 33308. Some of the papers included in the volume that are of special interest are listed separately below (refs. 26–37).

26. "Meteorological Summary of Hurricane Hugo," by James W. Brennan, pp. 1–12.

27. "Post-Hurricane Hugo Nearshore Side Scan Sonar Survey: Myrtle Beach to Folly Beach, South Carolina," by P. T. Gayes, pp. 95–112.

28. "Nearshore Profile Response Caused by Hurricane Hugo," by W. A. Birkemeyer et al., pp. 113–128.

29. "Effects of Hurricane Hugo on the South Carolina Coast," by D. K. Stauble et al., pp. 129–162.

30. "Factors Effecting Beach Morphology Changes Caused by Hurricane Hugo, Northern South Carolina," by Donald D. Nelson, pp. 163–180.

31. "Quantitative Evaluation of Coastal Geomorphic Changes in South Carolina after Hurricane Hugo," by E. Robert Thieler and Robert S. Young, pp. 187–200.

32. "Effects of Hurricane Hugo Storm Surge in Coastal South Carolina," by N. K. Coch and M. P. Wolf, pp. 201–228.

33. "Beach Scraping in North Carolina with Special Reference to Its Effectiveness during Hurricane Hugo," by John T. Wells and Jesse McNinch, pp. 249–263.

34. "The Effects of Hurricane Hugo on the Isle of Palms, South Carolina from Destruction to Recovery," by Michael P. Katuna, pp. 263–275.

35. "Impact of Hurricane Hugo on the South Carolina Coastal Plain Forest," by Donald D. Hook, Marilyn A. Buford, and Thomas M. Williams, pp. 291–300.

36. "Ecological Impact of Hurricane Hugo—Salinization of a Coastal Forest," by L. R. Gardner et al., pp. 301–318.

37. "Nature and Causes of Hurricane-Induced Ebb Scour Channels on a Developed Shoreline," by G. Lennon, pp. 237–248.

38. *Hurricane Hugo, September 10–22, 1989,* is a Natural Disaster Survey Report prepared and published by the National Weather Service branch of NOAA. The 61-page report has three appendixes and chapters on the meteorology of Hugo; a summary of preparedness actions, information, and warning services; an evaluation of the processing, interpretation, and dissemination of National Weather Service information; public response and user benefits; and findings and recommendations for both Puerto Rico and South Caro-

lina. Available from NOAA.

39. "Damage to Buildings: Hurricane Hugo," by S. M. Rogers, Jr., and P. R. Sparks, 1990, is a short paper that appeared in the October 1990 issue of *Shore and Beach* (vol. 58, no. 4).

40. "Water and Erosion Damage to Coastal Structures—South Carolina Coast, Hurricane Hugo, 1989," by H. Wang, *Shore and Beach,* vol. 58, no. 4 (1990), pp. 37–47.

41. "Surface Wind Speeds and Property Damage," by Richard D. Marshall, 1994, is part of a report published by the National Academy of Sciences. The NAS has teams of experts ready to visit sites of natural disasters immediately after they occur to record time-sensitive information and provide guidance on how to avoid such disasters in the future. The volume on Hurricane Hugo is titled *Hurricane Hugo: Puerto Rico, the Virgin Islands, and Charleston, South Carolina, September 17–22, 1989.* It is part of a series of reports called Natural Disasters Studies, an Investigative Series of the Committee on Natural Disasters. The Hugo report is volume 6 in that series. Available from National Academy Press, 2101 Constitution Avenue, NW, Box 285, Washington, DC 20055, (800) 624-6242 or (202) 334-3313.

Northeasters (Winter Storms)

42. "Nor'easters," by Robert E. Davis and Robert Dolan, is one the best and most thor-

ough treatments of winter storms available. It includes a scale for ranking these storms patterned somewhat after the Saffir/Simpson scale for hurricanes. Information on storm formation and tracking is included, along with good historical accounts. Published in *American Scientist,* vol. 81 (1993), pp. 428–439.

43. "New Respect for Nor'easters," by Ben Watson, was published in *Weatherwise,* vol. 46, no. 6 (December 1993), pp. 18–23. This article summarizes new research into the genesis and impacts of winter storms.

Barrier Islands and Beaches

General

44. *Barrier Islands: Process and Management,* edited by D. K. Stauble and O. T. Magoon, 1989, is a set of papers presented at Coastal Zone '89. The 327-page collection was published by the American Society of Civil Engineers in New York and should be of interest to the student of barrier islands. Papers pertinent to South Carolina include the following (refs. 45–48):

45. "Management of Drumstick Barrier Islands," by R. A. Davis, Jr., pp. 1–16.

46. "The South Carolina Coast: I—Natural Processes and Erosion; II—Development and Beach Management," by T. W. Kana, pp. 265–283.

47. "Shoreline Change along the South Carolina Coast," by F. J. Anders and D. W. Reed, pp. 296–310.

48. "Coastal Processes and Management Issues on Folly Island, South Carolina: 1850–1988," by Mark Hanson and David Harris, pp. 311–324.

49. *Beach Erosion in South Carolina,* by Timothy W. Kana, 1988. One of the most concise summaries of the state's beach erosion problems available, this 55-page booklet presents a view of the complex causes of erosion, the role of tidal deltas, the percentage of ocean shoreline subject to development (105 of 181 miles, or 58 percent), and management strategies. This pre-Hugo report suggests that the erosion problem is probably not as serious as believed, favors soft stabilization as the best mitigation approach, and recommends a 50-year perspective as a basis for management strategy, which may be short-sighted. Recommended reading for barrier island property owners, and available as SCSG-SP-88-1 report from the South Carolina Sea Grant Consortium.

50. *Using Common Sense to Protect the Coasts,* by Michael Weber, 1990. This brief document (24 pp.) contains basic information on the geology and ecology of barrier islands, the destructive effects of development on these areas, and legislation for their protection and management. Produced and distributed by The Coast Alliance (218 D Street, SE, Washington, DC 20003).

51. *Understanding Our Coastal Environ-ment,* edited by Ann Townsend Adkins, reprinted in 1988. This 40-page booklet prepared by the South Carolina Coastal Council provides an introduction to the varied and plentiful coastal environments in South Carolina, from barrier island beaches to oyster reefs, mudflats, and savannahs. Written for the layperson and illustrated with spectacular aerial photographs and understandable diagrams. Available from DHEC-OCRM.

52. *Ecological Characterization of the Sea Island Coastal Region of South Carolina and Georgia Resource Atlas,* edited by Jane Davis and others, 1980. This excellent multiauthor collection of oversized maps, charts, diagrams, and photographs tells you just about everything you could want to know about the sea island region (ecology, physiography, geology, climatology, and cultural and natural resources). Prepared for the EPA and the U.S. Fish and Wildlife Service, this report (FWS/OBS-79/43) was produced through the Biological Services Program and the Interagency Energy-Environment Research and Development Program.

53. *Atlantic Beaches,* by J. N. Leonard, 1972, presents the aesthetics of U.S. Atlantic coast beaches in words and pictures. Published as part of the American Wilderness Series by Time-Life Books, Rockefeller Center, New York, NY 10020.

54. *The Beaches Are Moving: The Drowning of America's Shoreline,* by Wallace Kaufman and Orrin Pilkey, 1979. This highly

readable account of the state of America's coastline explains the natural processes at work at the beach, provides a historical perspective of man's relation to the shore, and offers practical advice on how to live in harmony with the coastal environment. Published by Duke University Press, Box 90660, Durham, NC 27708-0660.

55. *Barrier Island Handbook,* by Steve Leatherman, 1979, is a nontechnical, easy to read paperback about barrier island dynamics and coastal hazards. Many of the examples are from the Maryland and New England coasts but are applicable to South Carolina as well. Available from Steve Leatherman, Department of Geography, Room 113, Social Science Building, University of Maryland, College Park 20742.

56. *At the Sea's Edge: An Introduction to Coastal Oceanography for the Amateur Naturalist,* by W. T. Fox, 1983. Excellent nontechnical, richly illustrated introduction to coastal processes, meteorology, environments, and ecology published by Prentice-Hall, Englewood Cliffs, NJ 07632.

57. *Dune/Beach Interaction,* edited by Norbert P. Psuty, 1988, is Special Publication 3 of the *Journal of Coastal Research,* published by the CERF. It is a compilation of 22 papers on dune-beach interactions as looked at by engineers, geologists, and geomorphologists from around the world, and is the proceedings of a special session sponsored by the Commission on the Coastal Environment, International Geographical Union, in conjunction with the Association of American Geographers, Portland, OR, April 1987.

58. *Living with the South Carolina Shore,* by William J. Neal, W. Carlyle Blakeney, Jr., Orrin H. Pilkey, Jr., and Orrin H. Pilkey, Sr., 1984, is the book updated by this new version. Out of print but probably still available at university libraries.

Geology

59. *Historical Inlet Atlas for South Carolina,* by Gary Zarillo and Larry Ward, 1983. This Sea Grant report reviews the history of South Carolina's major inlets, as determined from a survey of historical maps and aerial photographs, and describes their positions and general migration patterns. Recommended reading for anyone who owns or is considering buying property in the vicinity of an inlet. Available from South Carolina Sea Grant Consortium.

60. *Waves and Beaches,* by Willard Bascom, 1964, discusses beaches and coastal processes. Based on World War II research to assist amphibious landings, this classic may be the original coastal text. Updated periodically, it is a "must read" for beginners. Published by Anchor Books/Doubleday, Garden City, NY 11530.

61. *Beaches and Coasts,* 2nd edition, by C. A. M. King, 1972, is a classic treatment of beach and coastal processes. Published by St. Martin's Press, 175 Fifth Avenue, New York, NY 10010.

62. *Beach Processes and Sedimentation,* by Paul Komar, 1976, is the most up-to-date technical explanation of beaches and beach processes. Recommended only for serious students of the beach. Published by Prentice-Hall, Englewood Cliffs, NJ 07632.

63. *Terrigenous Clastic Depositional Environments,* edited by Miles Hayes and Tim Kana, 1976. Although compiled for a professional field course, this text is an excellent, detailed treatment of coastal sedimentary environments that makes good reading for the interested nonscientist. The numerous photographs and diagrams support the text's description of depositional systems in rivers, dunes, deltas, tidal flats and inlets, salt marshes, barrier islands, and beaches. Most of the examples are from South Carolina. Out of print, but available in university libraries.

64. *Barrier Islands from the Gulf of St. Lawrence to the Gulf of Mexico,* edited by Steve Leatherman, 1979. This collection of technical papers presents some of the current geological research on barrier islands. Of particular interest to students of South Carolina barrier islands is the lead paper, by Miles Hayes, "Barrier Island Morphology as a Function of Tidal and Wave Regime." Published by Academic Press, New York, NY.

65. *Report on the Geology of South Carolina,* by M. Tuomey, 1848. This old description of the coast describes landward-migrating

island shorelines, shoreline erosion, migrating dunes that buried trees and even houses, and the stabilizing effect of vegetation. More than a century later, some of these natural processes are still looked upon as new and unusual events. Published by A. S. Johnston, Columbia, SC.

66. *Coastal Environments: An Introduction to the Physical, Ecological and Cultural Systems of Coastlines,* by R. W. G. Carter, 1988, is an excellent text for almost all aspects of the coastal zone, although management of coastal environments is its emphasis. Published by Academic Press, New York, NY.

67. *Coasts: An Introduction to Coastal Geomorphology,* by Eric C. F. Bird, 3rd edition, 1984, is a good introduction to coastal types and classifications. It discusses tides, waves, and currents; changing levels of the sea; cliffed coasts; beaches, spits, and barriers; coastal dunes; estuaries and lagoons; deltas; and coral reefs and atolls. There is also a chapter on classification of coastal landforms. Published by Basil Blackwell of London.

68. *An Illustrated History of Tidal Inlet Changes in South Carolina,* by Gary A. Zarillo, Larry G. Ward, and Miles O. Hayes, 1985. This 76-page atlas is a well-illustrated history of inlet dynamics that includes background information on shoreline classification and tidal inlet morphology and processes. This is an update of ref. 59. Published in Charleston by the South Carolina Sea Grant Consortium.

69. *The Economic Impact of Proposed Coastal Setback and Renourishment Legislation on South Carolina* is a report prepared by Caroline D. Strobel and Douglas P. Woodward, 1988, for the South Carolina Coastal Council (now DHEC-OCRM) and the South Carolina Department of Parks, Recreation, and Tourism. Available from the OCRM or the Division of Research, College of Business Administration, University of South Carolina, Columbia 29208.

70. *Shoreline Change Maps, Cooperative Shoreline Movement Study,* 1984, available from NOAA, is a detailed study of shoreline change by the National Ocean Service, NOAA, and USACOE.

Island Environments

71. "Barrier Beaches of the East Coast," by P. J. Godfrey, is a well-written paper describing beaches and associated barrier island environments as related through the island processes. Published in *Oceanus,* vol. 19, no. 5 (1976), pp. 27–40.

72. *Know Your Mud, Sand, and Water: A Practical Guide to Coastal Development,* by K. M. Jurgensen, 1976, is a pamphlet describing the various island environments relative to development. Clearly and simply written. Recommended to island dwellers. Available from UNC Sea Grant.

73. *An Inventory of South Carolina's Coastal Marshes,* by R. W. Tiner, Jr., 1977, is Technical Report 23 (33 pp.), published by the South Carolina Marine Resources Center, Wildlife and Marine Resources Department, Charleston, SC 29412.

Recreation

74. *Recreation in the Coastal Zone,* 1975, is a collection of papers presented at a U.S. Department of the Interior, Bureau of Outdoor Recreation, Southeast Region, symposium. Outlines different views of recreation in the coastal zone and the approaches taken by some states to address recreation problems. The symposium was cosponsored by NOAA-OCRM, and the report is available from that office.

75. *Sea Islands of the South,* by Diana Gleasner and Bill Gleasner, 1980, is an excellent visitor's guide to the southeastern coast of the United States; it includes both nature information and descriptions of developed barrier islands. You will find descriptions, explanations, and identifications of everything from dunes and tides to birds and shells; also a guide to visitor information centers, accommodations, activities, and sightseeing points of interest. This guide is especially handy for the first-time traveler through North Carolina down to Florida, but it may be of interest to natives, too. Published by East Woods Press, 820 East Boulevard, Charlotte, NC 28203.

76. *South Carolina Public Beach & Coastal Access Guide,* by the South Carolina Department of Parks, Recreation, and Tourism and

South Carolina Coastal Council, provides location and site information on public and commercial outdoor recreation land, activities, and facilities for each county along the South Carolina coast. Digital maps augment detailed listings of facilities and accessibility; costs and phone numbers for more information are given where available. Updated periodically; available from South Carolina Department of Parks, Recreation, and Tourism.

77. *South Carolina Travel Guide: Smiling Faces, Beautiful Places,* by the South Carolina Department of Parks, Recreation, and Tourism, 1995. This book devotes three chapters to the coastal areas of South Carolina, providing some state history and highlighting recreational opportunities, accommodations, and tourist attractions for each area. Illustrated with color photographs. Available from the South Carolina Department of Parks, Recreation, and Tourism.

Shoreline Engineering and Beach Replenishment

78. *A Study of Shore Erosion Management Issues in South Carolina,* by J. B. London and others, 1981, is a good introduction to the erosion problems facing South Carolina and the possible solutions, from regulating land use to shoreline engineering. Case studies include Myrtle Beach, Pawleys Island, Hunting Island State Park, and Hilton Head Island. Recommended reading. Available from South Carolina Sea Grant Consortium as Technical Report SC-SG-81-1.

79. "Beach Nourishment along the Southeast Atlantic and Gulf Coasts," by Todd Walton and James Purpura, in *Shore and Beach* (vol. 45, July 1977), pp. 10–18. This article examines successes and failures of several beach nourishment projects, including the rapid postnourishment losses of beach fill at Hunting Island.

80. "Beach Behavior in the Vicinity of Groins," by C. H. Everts, 1979, is an interesting description of the effects of two groin fields in New Jersey. It concludes that groins deflect the movement of sand seaward, causing erosion in the downdrift shadow area, even if the groin compartments are filled with sand. Published in the *Proceedings of the Specialty Conference on Coastal Structures 79,* pp. 853–867; available as reprint 79-3 from the CERC.

81. *Low Cost Shore Protection,* by USACOE, 1981. This set of four reports written for the layman includes the introductory report, a property owner's guide, a guide for local government officials, and a guide for engineers and contractors. The reports summarize the Shoreline Erosion Control Demonstration Program and suggest a wide range of engineering devices and techniques to stabilize shorelines, including beach nourishment and vegetation. You should keep in mind that no erosion control devices are permitted in South Carolina. The reports are available from Section 54 Program, USACOE (DAEN-CWP-F), Washington, DC 20314.

82. *Shore Protection Guidelines,* by USACOE, 1971. Summary of the effects of waves, tides, and winds on beaches, and engineering structures used for beach stabilization. Available from the USACOE, Washington, DC.

83. *Publications List, Coastal Engineering Research Center (CERC) and Beach Erosion Board (BEB),* by USACOE. A bibliography (updated periodically) of published research by the Corps of Engineers. Free from the USACOE.

84. "Where Beaches Have Been Going: Into the Ocean," is by Gary Soucie, a former Outer Banks newsman who writes candidly about structural engineering devices, referring to their use in shoreline stabilization as an "utter failure." Published in *Smithsonian,* vol. 4, no. 3 (1973), pp. 54–61.

85. "An Analysis of Replenished Beach Design Parameters on U.S. East Coast Barrier Islands," by Lynn Leonard, Tonya Clayton, and Orrin Pilkey, is a scientific paper published in the *Journal of Coastal Research,* vol. 6, no. 1 (1990), pp. 15–36. The authors conclude that replenished beaches north of Florida generally have life spans of less than five years (storm frequency is a major factor) and document overestimates of beach life by the USACOE.

86. "A 'Thumbnail Method' for Beach Communities: Estimation of Long-Term Beach Replenishment Requirements," by Orrin H. Pilkey. This short paper, with tables and illus-

trations, demonstrates that current methods of estimating long-term volume requirements for replenished beaches are inadequate. The long-term volume required can be estimated by assuming that the initial restoration volume must be replaced at prescribed intervals depending on the geographic location. Applicable to any replenishment situation. Published in *Shore and Beach,* vol. 56, no. 3 (July 1988), pp. 23–31.

87. "Seawalls versus Beaches," by Orrin H. Pilkey and Howard L. Wright III, 1988, discusses how seawalls act to narrow the beaches in front of them. Clearly demonstrates that the problem is not *whether* seawalls negatively affect beaches, but *how* it happens. Published in *Journal of Coastal Research,* Special Issue 4, *The Effects of Seawalls on the Beach,* edited by N. C. Kraus and O. H. Pilkey, pp. 41–64.

88. *Shore Protection Manual,* by USACOE, 4th edition, 1984. This is the bible of shoreline engineering. It outlines the various types of engineering structures, including their destructive effects. Published in three volumes and for sale from the U.S. Government Printing Office; ask for stock no. 008-022-00218-9. Slated to be updated occasionally.

89. *Construction Materials for Coastal Structures,* by Moffatt and Nichol, Engineers, 1983. Lengthy (427 pp.) summary of the characteristics of a wide range of materials used in coastal structures, beach protection devices, and erosion control. This technical reference

guide should particularly interest coastal engineers and construction contractors who build such structures. Available from the CERC as Special Report 10.

90. "Structural Methods for Controlling Erosion," by C. R. O'Neil, 1986. Written for the layman, this short paper describes in very easy to understand terms the difference between each method's applicability to various coastal types and its compatibility with different forms of recreation. Available from the Cornell University Cooperative Extension Service as Information Bulletin 200, Ithaca, NY 14850.

91. *Conserving South Carolina Beaches through the 1990s: A Case for Beach Replenishment,* by Timothy W. Kana, 1990, is a 30-page booklet which makes a case for beach nourishment as the best beach stabilization tool available and the preferred alternative to armoring. It reviews other alternatives as well and provides cost analyses for various beach replenishment scenarios. The book doesn't address the down side of committing to beach nourishment (e.g., the true long-term cost to the community in terms of expense, the implications of holding the beach in place, the false sense of security that may be generated, environmental impacts, and dashed expectations); nevertheless, any community considering alternative ways of responding to beach loss should review this report. Available from DHEC-OCRM.

General

92. *Living by the Rules of the Sea,* by David M. Bush, Orrin H. Pilkey, and William J. Neal, 1996. This book is designed to guide the individual, manager, planner, architect, or government official in assessing a physical setting for coastal hazards, on a regional scale or an individual site. The book also addresses methods of mitigating storm damage to structures once the area is developed. Published by Duke University Press, Durham, NC 27708-0660.

93. "Coastal Hazard Mapping and Risk Assessment," by David M. Bush, 1993, presents a new technique for assessing the risk for property damage in coastal areas. This paper was presented at the 1993 Annual Forum of the National Committee on Property Insurance (now the Insurance Institute for Property Loss Reduction) in San Francisco. The proceedings from that conference were published as *Natural Disasters, Local and Global Perspectives.* Available from the Insurance Institute for Property Loss Reduction, 73 Tremont Street, Suite 510, Boston, MA 02108-3910.

94. *Citizen's Guide to Geologic Hazards,* by Edward B. Nuhfer and others, 1993. Written for the general public, this book discusses geologic hazards in understandable terms. It covers reactive minerals, asbestos, earthquakes, volcanoes, gas hazards, landslides, avalanches,

subsidence, floods, and coastal hazards. Available from the American Institute of Professional Geologists, 7828 Vance Drive, Suite 103, Arvada, CO 80003-2124, (303) 431-0831.

95. *Natural Hazards,* by E. A. Bryant, 1991, is a well-illustrated 294-page book that discusses many aspects of natural hazards. Major sections cover climatic hazards (storms, wind, oceanographic hazards, drought, flood, precipitation, fire); geological hazards (earthquakes, volcanoes, tsunamis, land instability); and the social impacts of these hazards. Published by Cambridge University Press, Cambridge, England.

96. *Coastal Hazards: Perception, Susceptibility, and Mitigation,* edited by Charles W. Finkl, Jr., 1995, is *Journal of Coastal Research* Special Issue 12. It covers coastal hazard issues such as hazard recognition and evaluation, sea level rise, storms, tsunamis, effects of humans on coastal environments, effects of coastal hazards on natural features, and hazard mitigation. Published by the CERF. The following two papers are contained in that volume:

97. "Hurricane Hazards along the Northeastern Atlantic Coast of the United States," by Nicholas K. Coch, pp. 115–147, treats the hurricane hazard in an area we don't normally think of as hurricane prone.

98. "Mitigation of Hurricane Property Damage on Barrier Islands: A Geological View," by David M. Bush and Orrin H. Pilkey, pp. 311–326, discusses coastal hazards and suggests ways to reduce property damage with low-cost, nonengineering approaches.

99. *Wide Awake Landing, Coastal Processes Workshop,* produced by the Research Planning Institute, Inc., 1981. Workshop proceedings with several articles on the erosion problem in South Carolina, including a reprint of "Development of Kiawah Island, South Carolina," by Miles Hayes. Privately published by RPI, Inc., Columbia, SC 29201.

100. *Natural Hazard Management in Coastal Areas,* by G. F. White and others, 1976, summarizes coastal hazards along the entire coast of the United States, discusses adjustments to such hazards and hazard-related federal policy and programs, and summarizes hazard management and coastal land-planning programs in each state. The appendixes include a directory of agencies, an annotated bibliography, and information on hurricanes. An invaluable reference, recommended to developers, planners, and managers. Available from NOAA-OCRM.

101. *The Risk of Hurricane Wind Damage to Buildings in South Carolina,* by P. R. Sparks, 1988. This 20-page pre-Hugo report on wind damage potential and building codes and their enforcement was prophetic. Read from the perspective of Hugo's destruction, property owners, community officials, developers, and builders should be making every effort to put the recommendations (on p. 17) of this report into action. Available from the South Carolina Sea Grant Consortium.

Earthquakes

102. *Studies Related to the Charleston, South Carolina, Earthquake of 1886—A Preliminary Report,* edited by D. W. Rankin, 1977, is a collection of technical papers analyzing the August 31, 1886, earthquake that struck the Charleston area, killing 60 people and causing extensive damage. The earthquake is thought to have been as strong as the 1971 earthquake in California's San Fernando Valley; the 1886 quake was felt as far away as Chicago. Available as USGS Professional Paper 1028 from the U.S. Government Printing Office.

103. "Earthquake History of South Carolina," by C. A. von Hake, in the *Earthquake Information Bulletin,* vol. 8, no. 6 (1976), pp. 34–38.

104. *South Carolina Earthquakes.* This pamphlet published by the Charleston Southern University Earthquake Education Center and the South Carolina Emergency Preparedness Division discusses the causes of and risks associated with earthquakes, specifically addressing concerns about future South Carolina quakes. Tells about the frequency of earthquakes in South Carolina and describes what to expect during the event and what steps to take afterward. Available from the Earthquake Education Center, Charleston Southern University, Charleston, SC 29423; or South Carolina Emergency Preparedness Division, Columbia 29201.

105. "Identification of a Northwest Trending Seismographic Graben Near Charleston, South Carolina," by Gered Lennon, 1986, explains the results of studies conducted in the area of the 1886 Charleston earthquake that seem to suggest the existence of two faults extending northwest in the Charleston area. Master's thesis, Department of Geology, University of South Carolina, Columbia.

106. *A Review of Earthquake Research Application in the National Earthquake Damage Reduction Program* (NERP) *1977–1987,* edited by Walter Hays, 1988. USGS Open-File Report OF-88-0013-A.

107. *Catastrophic Earthquakes: The Need to Insure Against Economic Disaster,* by the National Committee on Property Insurance, 1989. From the Earthquake Project, a coalition of insurance companies and their trade associations, this 120-page report (plus 188 pages of appendixes) addresses the national threat of a catastrophic earthquake's impact on the U.S. economy and the insurance industry's role in preparation and response. Discusses the 1886 Charleston quake (estimated magnitude 6.5–7.0) and other central and eastern U.S. events. No matter where the "big one" occurs, the projected impact on the national and local economies is sobering. Prepared by the National Committee on Property Insurance (now the Insurance Institute for Property Loss Reduction). Available from the Insurance Institute for Property Loss Reduction, 73 Tremont Street, Suite 510, Boston, MA 02108-3910.

108. "Internally Consistent Pattern of Seismicity Near Charleston, South Carolina," by P. Talwani, gives information on the Charleston seismic setting. For the serious student of earthquakes. Published in the professional journal *Geology,* vol. 10 (1982), pp. 654–658.

109. "Holocene Neotectonic Distortion in the Charleston, S.C. Region," by D. J. Colquhoun, 1984, also gives detailed information on the Charleston seismic setting. For the serious student or researcher. Published in the Final Technical Report to the U.S. Geological Survey, Menlo Park, CA, 22 pp.

110. "The Quake That Swallowed a City," by George R. Clark, tells the story of the Port Royal, Jamaica, earthquake of June 7, 1692. More than 2,500 people died, many "swallowed up in graves of quicksand." An interesting mix of old accounts and new perspectives of earthquake hazards. Published in *Earth* magazine (April 1995), pp. 34–38.

111. "Recent Vertical Crustal Movements in the South Carolina Coastal Plain: Implications for Neotectonic Activity," by C. M. Poley, 1982. This master's thesis from the Department of Geology, University of South Carolina, Columbia, examines the results of periodic surveys of the coastal plains by the South Carolina Geodetic Survey, which include evidence of downwarping in the Charleston area and uplift in the area of Kiawah Island associated with movement along the Ashley River Fault.

112. *Earthquake Resistant Building Design and Construction,* by Norman Green, was published in 1987 by Elsevier, New York. This is a clear and concise treatise on the mitigation of earthquake hazards at the design and construction levels for builders, contractors, and code enforcement personnel.

Vegetation

Remember, in South Carolina a permit is required for planting on oceanfront dunes.

113. *Building and Stabilizing Coastal Dunes with Vegetation* (UNC-SG-82-05) and *Planting Marsh Grasses for Erosion Control* (UNC-SG-81-09), by S. W. Broome, W. W. Woodhouse, Jr., and E. D. Seneca, 1982. These publications on using vegetation as stabilizers are available from UNC Sea Grant.

114. *Seacoast Plants of the Carolinas for Conservation and Beautification,* by K. E. Braetz, 1973, is an excellent discussion of beach and dune environments with respect to natural plants. Suggests plants to stabilize and protect dunes and landscape the beach, and evaluates perennial beach plants. Includes descriptions and illustrations of various natural and ornamental plants. Available from coastal offices of the District Conservationists or from UNC Sea Grant.

115. *The Dune Book: How to Plant Grasses for Dune Stabilization,* by Johanna Seltz, 1976. This brochure outlines the importance of sand dunes and means of stabilizing them

through grass plantings. Available from UNC Sea Grant.

116. *Vegetative Dune Stabilization in North Carolina,* by Carl T. Blake and others, 1973, is a one-page agronomy information leaflet that outlines types of plants used for dune stabilization. Includes instructions on transplanting. Available from Agricultural Information, North Carolina State University, Raleigh, NC 27607.

117. *Artificial Seaweed for Shoreline Erosion Control?* by Spencer Rogers, Jr., 1986, is an excellent summary of worldwide attempts to use artificial seaweed to protect beaches. Concludes that artificial seaweed is not a very effective means of erosion control. Available from UNC Sea Grant as publication UNC-SG-WP-86-4.

Site Analysis

118. "Beach Erosion Trends along the South Carolina Coast," by D. K. Hubbard and others, 1977, outlines South Carolina's coastal types and rates of erosion. The overall conclusion is that the areas south of Winyah Bay are unstable, but areas backed by beach ridges have lower short-term erosion rates. Published in *Coastal Sediments '77,* the proceedings of the Fifth Symposium of the Waterways, Port, Coastal and Ocean Division, American Society of Civil Engineers, held at Charleston in November 1977, pp. 797–814. This volume of technical papers on coastal problems includes several other papers relevant to the South Carolina coast. *Note:* the erosion rates are dated. For current erosion rates contact the OCRM.

119. *Beach Erosion in South Carolina,* by Miles O. Hayes, Thomas F. Moslow, and Dennis K. Hubbard, 1984 and 1978, characterizes beach erosion trends along the South Carolina shore based on beach profiles, beach processes, and erosional-depositional history. Also contains the original classification of the South Carolina shore into four coastal types. This 99-page report was printed by the Department of Geology, University of South Carolina, Columbia, and should be available at or through university libraries.

120. *Modern Clastic Depositional Environments: South Carolina, Charleston to Columbia, South Carolina,* by Miles O. Hayes, and Walter J. Sexton, is a field trip guidebook published by the American Geophysical Union. It was prepared for the Twenty-eighth International Geological Congress held in Washington, D.C., in 1989. The field trip was run on July 20–25, 1989, and the guidebook (T371) is available from the American Geophysical Union, 2000 Florida Avenue, NW, Washington, DC 20009; (202) 462-6900.

121. Erosion Rate Maps for South Carolina. These are official state erosion rates on file and available from DHEC-OCRM. They are updated every several years in accordance with the Beachfront Management Act.

Water Resources

122. *The Occurrence, Availability, and Chemical Quality of Ground Water, Grand Strand Area and Surrounding Parts of Horry and Georgetown Counties, South Carolina,* by Allen Zack, 1977, evaluates geologic aquifers in the area and problems relating to groundwater availability and quality. South Carolina Water Resources Commission Report 8. Available from the Water Resources Division of DHEC.

123. *Ground Water in the Coastal Plains Region, a Status Report and Handbook,* compiled by A. D. Park, 1979. This report was prepared for the Coastal Plains Regional Commission and addresses the subject of groundwater in the five southeastern states. Although no specific South Carolina coastal groundwater problems are presented, there is a good summary of the state's programs and needs. Other states' problems and programs provide a basis for comparison. There is an extensive list of references for South Carolina, including studies on saltwater intrusion, aquifer depletion, and water contamination associated with waste disposal. The appendix provides a list of groundwater agencies for the states. Available from Coastal Plains Regional Commission, 215 East Bay Street, Charleston, SC 29401.

124. *Ground Water: The Crisis Below,* by Betsy Neal, Kay Jackson, and Shannon Lowry, compiles a series of articles that appeared in the *Savannah Morning News* from October 26

through October 31, 1981, focusing on the growing problem of adequate water supplies in the Southeast. The book includes a particularly interesting article titled "Low Country Is Buying Time," which points out water problems on South Carolina's barrier islands, particularly Hilton Head. Available from the *Savannah News-Press*, P.O. Box 1088, Savannah, GA 31402.

125. *Water Resources Development in South Carolina*, by USACOE, 1973. Descriptions include coastal projects at Hunting Island Beach State Park, Murrells Inlet, and half a dozen smaller coastal projects. This 58-page report should be available through libraries of the district offices of the USACOE.

126. *Your Home Septic System, Success or Failure* is a brochure with answers to commonly asked questions on home septic systems. Lists agencies that supply information on septic tank installation and operation. Available from UNC Sea Grant.

127. *Report of Investigation of the Environmental Effects of Private Waterfront Lands*, by W. Barada and W. M. Partington, 1972. An enlightening reference that addresses the effects of finger canals on water quality. Available from the Environmental Information Center, Florida Conservation Foundation, Inc., 935 Orange Avenue, Winter Park, FL 32789.

128. *National Water Summary*, an annual report by the USGS Division of Water Resources, with state-by-state summaries. The reports detail the history of water development,

use, and management. Available through the USGS Division of Water Resources.

Individual Islands

129. *Coastal and Fluvial Landforms: Horry and Marion Counties, South Carolina*, by Bruce Thom, 1967, is a Ph.D. dissertation published by Louisiana State University Press, Baton Rouge, that traces the recent geologic history of the northern South Carolina coastal region. It mentions freshwater peat outcrops in the vicinity of 80th Avenue, North Myrtle Beach, and pine stumps on Myrtle Beach, which indicate an eroding shoreline in the mid-1960s.

130. *Debidue Beach and North Inlet Landform History since 1733*, by T. M. Williams and C. A. Gresham, 1976. Technical Paper 4, Department of Forestry, Clemson University, Clemson, SC 29631, 15 pp.

131. *Hydraulics and Dynamics of North Inlet, South Carolina, 1975–76*, by Dag Nummedal and Stan Humphries, 1978. This engineering report is a technical analysis of inlet dynamics and the relationship between ebb-dominated tidal channels and adjacent beaches. General Investigation of Tidal Inlets (GITI) Report 16, USACOE, CERC.

132. "Variations in Tidal Inlet Processes and Morphology in the Georgia Embayment," by Dennis Hubbard, 1977. Report based on a thesis study of tidal inlet variability. Available at the university library and Department of Geol-

ogy, University of South Carolina, Columbia.

133. *Time and Tide on Folly Beach, South Carolina (a History)*, by Gretchen Stringer-Robinson, is a 100-page booklet published in 1989 by the author that details the history of the development of Folly Beach. There are excellent stories of people and events over the past century. Available at local bookstores and hotel gift shops.

134. *Beaches and Barriers of the Central South Carolina Coast*, edited by Dag Nummedal, 1977, is a guidebook prepared for a field trip to Sullivans Island and the Isle of Palms in conjunction with Coastal Sediments '77 (see ref. 118). Although it is written in technical terms, nonscientists interested in the South Carolina coast will find this an informative collection of papers by several of the state's coastal experts. Available at university library and Department of Geology, University of South Carolina, Columbia.

135. *Folly Beach, South Carolina: Survey Report on Beach Erosion Control and Hurricane Protection*, by USACOE, 1979. This report with appendixes is a detailed account of the erosion problem at Folly Beach, the history of that erosion, earlier actions to combat erosion, and proposed plans for stabilizing the shoreline, including a plan for beach nourishment. Of particular interest are predicted 50-year shoreline positions one to two blocks deep into the island front. The report is available for inspection at the USACOE District Office in Charleston.

136. *Environmental Inventory of Kiawah Island,* by William Campbell and John Dean, 1975, is a study prepared for Coastal Shores, Inc., Environmental Research Center, Columbia, SC.

137. *Preliminary Design Engineering and Permit Application for Breaching Kiawah Spit North of Captain Sam's Inlet, Final Report to the Seabrook Island Company,* by T. W. Kana and others, 1981, describes an inventive way to replenish a beach by artificially relocating an inlet channel (a process that occurs naturally anyway), thus releasing tons of sediment to flow naturally to the next island. This 43-page report, produced by Research Planning Institute, Columbia, SC, is available for inspection at the Research Planning Institute and at DHEC-OCRM.

Coastal Management

138. "Hurricanes Gilbert and Hugo Send Powerful Messages for Coastal Development," by Edward Robert Thieler and David Michael Bush, 1991. This article, which appeared in the *Journal of Geological Education,* vol. 39 (1991), pp. 291–299, compares the characteristics and impacts of Hugo and Gilbert and discusses how the types and designs of buildings in Mexico and South Carolina contributed to the damage.

139. "Evolution of Coastal Hazards Policies in the United States," by Rutherford H. Platt, discusses coastal erosion history and the evolu-tion of the public response to erosion. It is a detailed treatment of how the United States and individual states have arrived at their current coastal zone management policies. It was published in the professional journal *Coastal Management,* vol. 22 (1994), pp. 265–284.

140. *State of South Carolina Coastal Management Program and Final Environmental Impact Statement,* by the Office of Coastal Zone Management, NOAA, and the South Carolina Coastal Council, 1979. This lengthy volume describes the South Carolina Coastal Program as approved under the federal coastal zone management program, including policies, practices, and background of the Coastal Council (now DHEC-OCRM), and areas over which it has authority. An excellent resource for anyone interested in the regulation of the South Carolina coastal zone. Available for inspection at DHEC-OCRM.

141. *Permitting Rules and Regulations,* by the South Carolina Coastal Council. This is the official rules and regulations for permitting in critical areas of the coastal zone as defined under Act 123 (1977) and administered by the OCRM. Available from DHEC-OCRM.

142. *Questions and Answers on the National Flood Insurance Program,* FEMA publication FIA-2, 1983 (updated March 1992), is a pamphlet explaining the basics of flood insurance and providing addresses of FEMA offices. Available from regional offices of FEMA.

143. *Development of the Coast: Facing the Tough Issues,* is the proceedings of a Coastal States Organization conference held in Charleston in 1979. It gives an abbreviated overview of the wide range of problems generated by coastal development. Available from CSO, Conference Management Associates, Ltd., 1044 National Press Building, Washington, DC 20045.

144. *Introduction to Coastal Management,* by Timothy Beatley, David J. Brower, and Anna K. A. Schwab, published in 1994 by Island Press of Washington, D.C. (210 pp.), is the first complete book on the growing topic of coastal management. It could be used as a text for college-level coastal management courses or as a reference book by anyone working or interested in coastal management.

145. *Who's Minding the Shore,* by the Natural Resources Defense Council, Inc., 1976, is a guide to public participation in the coastal zone management process. Defines coastal ecosystems and outlines the Coastal Zone Management Act, coastal development issues, and means of citizen participation in the management process. Lists sources of additional information. Available from NOAA's Office of Coastal Zone Management.

146. *Ecological Determinants of Coastal Area Management,* 2 volumes, by Francis Parker, David Brower, and others, 1976. Volume 1 defines natural processes that operate within coastal areas, outlines development's disturbing influences on island environments, and suggests management tools and techniques. Volume 2 is a set of appendixes with

information on coastal-ecological systems, human impacts on barrier islands, and tools and techniques for coastal area management. Also contains a good barrier island bibliography. Available from the Center for Urban and Regional Studies, University of North Carolina at Chapel Hill, 108 Battle Lane, Chapel Hill 27514.

147. *Coastal Environmental Management,* prepared by the Conservation Foundation, 1980, gives guidelines for conserving resources and protecting against storm hazards. Part 2 presents a complete list of federal agencies and their authority under the law to regulate coastal zone activities. A good reference for planners and persons interested in wise land management. Available from the U.S. Government Printing Office.

148. *Projecting Future Sea Level Rise: Methodology, Estimates to the Year 2100, & Research Needs,* by John Hoffman, Dale Keyes, and James Titus, 2nd edition, 1983. This classic report for the EPA estimates future sea level rises and includes chapters on the scientific basis for projecting rise, the technique used, sea level scenarios to the year 2100, impacts of sea level rise, and research needs. Available from the EPA or the U.S. Government Printing Office as EPA 230-09-007.

149. *Atmospheric Carbon Dioxide and the Greenhouse Effect,* 1989. This 36-page DOE pamphlet is divided into six sections: "Introduction," "Increases in Atmospheric Carbon Dioxide," "Climate," "Plants," "Sea Level,"

and "Response to the Challenge." An easy-reading summary of the state of current thought on greenhouse warming is written in the format of answers to 15 questions about the topic. Available from DOE, Office of Energy Research, Office of Basic Energy Sciences, Washington, DC 20545, as bulletin DOE/ER-0411.

150. *Greenhouse Effect and Sea Level Rise,* edited by M. C. Barth and J. G. Titus, 1984. This 325-page text treats the probable effects of the sea level rise. Of particular interest is Chapter 4: "The Physical Impact of Sea Level Rise in the Area of Charleston, South Carolina," by T. W. Kana, J. Michel, M. O. Hayes, and J. R. Jensen (p. 105–50); and Chapter 7: "Economic Analysis of Sea Level Rise: Method and Results," by M. J. Gibbs (p. 215–51), which includes the Charleston area as a case study. The rise will result in shoreline retreat, greater storm surge penetration, saltwater intrusion and aquifer loss, marsh loss, and increased riverine flooding. Property values will decline, damage to property will increase, and new environmental threats will emerge (e.g., flooding of old waste disposal sites). Published by Van Nostrand Reinhold, New York, NY.

151. *Projected Impact of Relative Sea Level Rise on the National Flood Insurance Program,* by FEMA, 1991, concludes that period mapping of coastal flood hazard areas is necessary to stay abreast of the impact of the sea level rise, but that elevation requirements of the present NFIP program provide at least a 20-

year cushion (to 2010) for study and adjustment of construction elevation requirements. This 72-page report includes projections of numbers of households in the coastal floodplains through the year 2100. Produced by FEMA, Federal Insurance Administration, Washington, DC.

152. *Managing Coastal Erosion,* by the Committee on Coastal Erosion Zone Management, for the National Research Council, 1990, is a 182-page report on coastal erosion and its management written by a blue ribbon panel of experts. Chapters include "Coastal Erosion: Its Causes, Effects, and Distribution"; "Management and Approaches"; "The National Flood Insurance Program"; "State Programs and Experiences"; and "Predicting Future Shoreline Changes." There are seven appendixes. Available from National Academy Press, 2101 Constitution Avenue, NW, Box 285, Washington, DC 20055.

153. *How to Use a Flood Map to Protect Your Property: A Guide for Interested Private Citizens, Property Owners, Community Officials, Lending Institutions, and Insurance Agents,* December 1994. This 22-page tabloid-sized publication helps readers understand Flood Insurance Rate Maps (FIRMs), which establish the extent of flood hazard within a flood-prone community. Available from FEMA as publication FEMA-258.

154. *Cities on the Beach: Management Issues of Developed Coastal Barriers,* edited by Rutherford Platt, Sheila Pelczarski, and Bar-

bara Burbank, 1987. The 28 papers in this book concern barrier island development. There are eight main sections: "Introduction," "The Coastal Barrier Resources Act," "Geographical Characteristics of Coastal Barriers," "Planning and Growth Management," "Shoreline Management," "Hazard Management," "Legal Issues," and "Management Alternatives." Available from the Department of Geography, University of Chicago, 5828 S University Avenue, Chicago, IL 60637-1583, as Research Paper 224.

155. *Ocean and Coastal Law,* by R. Hildreth and R. Johnson, 1983, discusses (1) problems posed by earlier nonmanagement of public resources and the gradual public awakening to the need for a comprehensive legal framework for the coastal zone, (2) ownership and boundary questions, (3) state common law, (4) offshore issues, (5) alteration of waterways and wetlands, and (6) federal and state coastal zone management programs. For anyone with a serious interest in marine legal issues. Published by Prentice-Hall, Englewood Cliffs, NJ 07632.

156. *South Carolina Blue Ribbon Committee on Beachfront Management,* 1987, reports the findings of the committee's investigation of beach erosion along the South Carolina shore, makes recommendations for how the beaches and dunes should best be managed for South Carolina citizens, and proposes long-term solutions to the problems identified.

157. *Catastrophic Coastal Storms,* by David R. Godschalk, David J. Brower, and Timothy Beatley, 1989, contains extensive information on mitigation and development management in at-risk coastal locations. Published by Duke University Press, Durham, NC 27708-0660.

158. *50 Years of Population Change along the Nation's Coasts 1960–2010,* by Thomas J. Culliton and others. This short (41-page) illustrated report is the second in NOAA's Coastal Trends Series, which looks at population trends along U.S. coastal areas. Available from NOAA.

Coastal Construction

Design

159. *Surviving the Storm: Building Codes, Compliance and the Mitigation of Hurricane Damage,* by the All-Industry Research Advisory Council (AIRAC), 1989. Available from AIRAC, Oak Brook, IL.

160. "Wind Conditions in Hurricane Hugo and Their Effect of Buildings in Coastal South Carolina," by P. R. Sparks, pp. 13–24, is a contribution to the *Journal of Coastal Research,* Special Issue 8, on Hurricane Hugo (see ref. 25). Property owners should read this paper carefully because the author gives a good evaluation of how and why buildings failed in Hurricane Hugo, and why the Standard Building Code's wind-resistance requirements were inadequate. Unfortunately, the

code and South Carolina's building requirements are still inadequate, and the next hurricane is on the horizon. The reference is available through university libraries.

161. "Hurricane Hugo and Implications for Design Professionals and Code Writing Authorities," by H. S. Saffir (of Saffir/Simpson scale fame), pp. 25–32, is another contribution to the *Journal of Coastal Research,* Special Issue 8, on Hurricane Hugo (see ref. 25) and a companion reference to the one above. Recommended reading for all property owners and community officials.

162. *Wind Forces on Structures,* 1961, is Paper 3269 of the American Society of Civil Engineers, 345 E 47th Street, New York, NY 10017-2398.

163. *Minimum Design Loads for Buildings and Other Structures,* American Society of Civil Engineers, 1994. Builders and design engineers should be familiar with this highly technical set of standards for construction. Replaces ANSI/ASCE 7-88. Available as publication ANSI/ASCE 7-93 from the American Society of Civil Engineers, 345 E 47th Street, New York, NY 10017-2398.

164. *The South Florida Building Code,* 1988 edition, Board of County Commissioners, Metropolitan Dade County, Miami, FL.

165. "A Study of the Effectiveness of Building Legislation in Improving the Wind Resistance of Residential Structures," by S. M. Rogers, Jr., P. R. Sparks, and K. M. Sparks,

1985. Published in *Proceedings of the Fifth U.S. National Conference on Wind Engineering,* a symposium held at Texas Tech University, Lubbock, November 5–8, 1985.

166. *Connectors for High Wind-Resistant Structures: Retrofit and New Construction,* 1992, Simpson Strong-Tie Company, Inc.

167. *Hurricane Resistant Construction Manual,* 1988, Southern Building Code Congress International, Inc., Birmingham, AL.

168. *Standard Building Code,* 1988, Southern Building Code Congress International, Inc., Birmingham, AL.

169. "The Winds of Change?" by Paul Tarricone, appeared in the professional journal *Civil Engineering* (vol. 64, no. 1, p. 42) in January 1994.

170. *Mitigating Damages in Hawaii's Hurricanes: A Perspective on Retrofit Options,* by George F. Wallace, 1994. This was written for the Insurance Institute for Property Loss Reduction and published as Occasional Paper 3.

171. *Building Practices for Disaster Mitigation,* edited by R. Wright, S. Kramer, and C. Culver, 1973. Prepared for the U.S. Department of Commerce, National Bureau of Standards (now the National Institute of Standards Technology) as NBS Building Series 46.

172. *Coastal Design: A Guide for Builders, Planners, and Homeowners,* by Orrin H. Pilkey, Sr., and others, 1983, is a companion volume and construction guide for books of the Living with the Shore series. Includes discussions of shoreline types, individual residence construction, making older structures stormworthy, high-rise buildings, mobile homes, coastal regulations, and the future of the coastal zone. Published by Van Nostrand Reinhold, New York, NY.

173. *Elevated Residential Structures,* prepared by the American Institute of Architects Foundation (1735 New York Avenue NW, Washington, DC 20006) for FEMA, 1984. An excellent outline of coastal and riverine flood hazards and the need for proper planning and construction. Discusses the National Flood Insurance Program, site analysis and design, design examples, and construction techniques. Includes illustrations, glossary, references, and worksheets for estimating building costs. Available as stock no. 1984-0-438-116 from the U.S. Government Printing Office; also available as FEMA-54 from FEMA offices.

174. *Flood Emergency and Residential Repair Handbook,* prepared by the National Association of Homebuilders Research Advisory Board of the NAS, 1980. Guide to floodproofing plus step-by-step cleanup procedures and repairs for household goods and appliances. Available from the U.S. Government Printing Office as stock no. 023-000-00552-2.

175. *Wind-Resistant Design Concepts for Residents,* by Delbart B. Ward. Vivid sketches and illustrations explain construction problems and methods of tying structures to the ground. Considerable text and excellent illustrations are devoted to methods for strengthening residences. Offers recommendations for relatively inexpensive modifications that will increase the safety of residences subject to severe winds. Available as TR-83 from the Civil Defense Preparedness Agency, Department of Defense, Pentagon, Washington, DC 20301; or from the Civil Defense Preparedness Agency, 2800 Eastern Boulevard, Baltimore, MD 21220.

176. *Interim Guidelines for Building Occupant Protection from Tornadoes and Extreme Winds,* TR-83A, and *Tornado Protection—Selecting and Designing Safe Areas in Buildings,* TR-83B, are supplements to reference 175 and are available from the same address.

177. *The Uniform Building Code* is one of the commonly applied building codes in the United States. Available from International Conference of Building Officials, 5360 S Workman Mill Road, Whittier, CA 90601.

178. *Structural Failures: Modes, Causes, Responsibilities,* 1973. See especially the chapter titled "Failure of Structures Due to Extreme Winds," pp. 49–77. Available from the Research Council on Performance of Structures, American Society of Civil Engineers, 345 E 47th Street, New York, NY 10017.

179. *Hurricane-Resistant Construction for Homes,* by Todd L. Walton, Jr., and Michael R. Barnett, 1991, gives guidelines for wood-frame, masonry, and brick construction; pole houses; and special considerations such as

roofs, doors, glass, shutters, and siding. Available as Bulletin 16 (MAP-16) from Florida Sea Grant College Program, University of Florida, Gainesville 32611.

180. *Guidelines for Beachfront Construction with Special Reference to the Coastal Construction Setback Line,* by C. A. Collier and others, 1977, Report 20 from Florida Sea Grant College Program, University of Florida, Gainesville 32611.

181. *Pole House Construction and Pole Building Design.* Available from the American Wood Preservers Institute, 1651 Old Meadows Road, McLean, VA 22101.

182. *Standard Details for One-Story Concrete Block Residences,* by the Masonry Institute of America. Contains nine fold-out drawings that illustrate the details of constructing a concrete block house. The principles of reinforcement and good connections this book presents are aimed at design for seismic zones but apply to hurricane zones as well. Written for both layman and designer. Available as Publication 701 from Masonry Institute of America, 250 Beverly Boulevard, Los Angeles, CA 90057.

183. *Building Performance: Hurricane Andrew in Florida; Observations, Recommendations, and Technical Guidance,* by FEMA, Federal Insurance Administration, 1992. This very well illustrated color booklet includes background information about the storm, on-site observations of the damaged area, and recom-mendations for reducing future damage. Available from FEMA; ask for publication FIA-22.

184. *Building Performance: Hurricane Iniki in Hawaii; Observations, Recommendations, and Technical Guidance,* by FEMA, Federal Insurance Administration, 1992, in cooperation with the Hawaii Office of Civil Defense and Kauai County. This well-illustrated color booklet contains the same type of information as reference 183. Available from FEMA as publication FIA-23.

185. *Building Construction on Shoreline Property* is a checklist by C. A. Collier. Homeowners and prospective buyers of coastal property will find this pamphlet a handy guide for evaluating location, elevation, building design and construction, utilities, and inspection. Available from either the Marine Advisory Program, G022 McCarty Hall, University of Florida, Gainesville 32611, or the Florida Department of Natural Resources, Bureau of Beaches and Shores, 202 Blount Street, Tallahassee 32304.

186. *Coastal Construction Manual,* prepared by Dames & Moore and Bliss & Nyitray, Inc., for FEMA, 1986. Guide to the coastal environment with recommendations on site and structure design relative to the National Flood Insurance Program. Includes discussions of the program, building codes, coastal environments, and examples. The second edition includes a chapter on the design of large structures at the coast. Appendixes include design tables, bracing examples, design worksheets, equations and procedures, construction costs, and a list of references. Also includes engineering computer program listings and a sample construction code for use by coastal municipalities. Available from the U.S. Government Printing Office as publication FEMA-55 and from FEMA offices.

187. *Design Guidelines for Flood Damage Reduction,* prepared by the American Institute of Architects for FEMA, 1981, to encourage appropriate design and construction practices in flood-prone areas. FEMA publication FEMA-15.

188. *Wind and the Built Environment: U.S. Needs in Wind Engineering and Hazard Mitigation,* report of the Panel on the Assessment of Wind Engineering Issues in the United States, for the Committee on Natural Disasters, NRC, 1993 (110 pp.). State-of-the-art report on wind hazards by a panel of experts. Chapters include "The Nature of Wind"; "Wind Engineering Research Needs"; "Mitigation, Preparedness, Response, and Recovery"; "Education and Technology Transfer"; and "Cooperative Efforts." Available from National Academy Press, 2101 Constitution Avenue, NW, Box 285, Washington, DC 20055.

189. *The Home Builder's Guide for Earthquake Design,* by the Applied Technology Council, 1980. This booklet is a condensed version of a 1978 document on methods for

seismic design and construction of single-family homes published by the Department of Housing and Urban Development's Office of Policy Development and Research. It provides information on site selection; foundations; roofs, floors, and walls; and construction details for building homes designed to withstand earthquakes. Available from the Department of Housing and Urban Development, Office of Policy Development and Research, Washington, DC.

190. *Mitigation of Flood and Erosion Damage to Residential Buildings in Coastal Areas,* by FEMA, 1994. This report on floodproofing investigations includes a couple of case studies from the Carolinas and is a good starting point to review floodproofing techniques; includes additional reading list. Available as publication FEMA-257 from FEMA.

191. *1991 Uniform Building Code with 1993 Supplement,* by International Conference of Building Officials, 5360 Workman Mill Road, Whittier, CA 90601-2298.

192. *Wood Frame House Construction,* revised edition, by L. O. Anderson, Craftsman Book Company, 6058 Corte del Cedro, Carlsbad, CA 92009.

193. *Wood Frame Construction Manual for One and Two Family Dwellings,* American Forest and Paper Association.

194. APA *Residential Construction Guide,* American Plywood Association, P.O. Box 11700, Tacoma, WA 98411-0700.

195. *Diaphragms*, American Plywood Association, Publication 1350, P.O. Box 11700, Tacoma, WA 98411-0700.

196. *Handbook to the Uniform Building Code,* by Vincent R. Bush, International Conference of Building Officials, 5360 Workman Mill Road, Whittier, CA 90601-2298.

Retrofitting

197. "Retrofit of Existing Buildings," by James L. Stratta, 1985, is a detailed and well-illustrated (with line drawings) article showing ways to retrofit several types of existing buildings to reduce potential damage from earthquakes. Not specifically for South Carolina, but the concepts apply here. Earthquake-resistant buildings also stand up well to hurricanes, as shown especially in Yucatán, Mexico, after Hurricane Gilbert. The article is in USGS Open-File Report 85-731, *A Workshop on Reducing Potential Losses from Earthquake Hazards in Puerto Rico,* edited by Walter W. Hays and Paula L. Gori, compiled by Carla J. Kitzmiller, pp. 124–151. Available from the USGS.

Mobile Homes

198. *Protecting Mobile Homes from High Winds,* prepared by the Civil Defense Preparedness Agency, 1974, as TR-75. This excellent 16-page booklet outlines methods of tying down mobile homes and means of protection such as positioning and windbreaks. Available from the U.S. Government Printing Office as publication 1974-0-537-785, or from the U.S. Army Publications Center, Civil Defense Branch, 280 Eastern Boulevard (Middle River), Baltimore, MD 21220.

199. *Suggested Technical Requirements for Mobile Home Tie Down Ordinances,* prepared by the Civil Defense Preparedness Agency, 1974, as TR-73-1, should be used in conjunction with TR-75 (ref. 198). Available from the U.S. Army Publication Center, Civil Defense Branch, 2800 Eastern Boulevard (Middle River), Baltimore, MD 21220.

200. *A Study of Reaction Forces on Mobile Home Foundations Caused by Wind and Flood Loads,* by Felix Y. Yokel and others, 1981. This technical report from the National Bureau of Standards (now the National Institute of Standards Technology) emphasizes that diagonal ties resist wind forces while vertical ties are more effective for resisting flood forces. It is NBS Building Science Series 132, available from the U.S. Government Printing Office.

201. *Manufactured Home Installation in Flood Hazard Areas,* prepared by the National Conference of States on Building Codes and Standards for FEMA, 1985, is a 110-page guide to design, installation, and general characteristics of manufactured homes with respect to coastal and flood hazards. Anyone contemplating buying a manufactured home or already living in one should read this publication and

follow its suggestions to lessen potential losses from flood, wind, and fire. Available from the U.S. Government Printing Office as publication 1985-529-684/31054; or from FEMA as publication FEMA-85.

Earthquake-Resistant Construction

202. *Buildings at Risk: Seismic Design Basics for Practicing Architects,* American Institute of Architects/Association of Collegiate Schools of Architecture (AIA/ACSA) Council on Architectural Research, 1725 New York Avenue, NW, Washington, DC 20006.

203. *The Home Builder's Guide for Earthquake Design,* by FEMA, 1992. Available from FEMA as FEMA-232.

204. *Reducing the Risks of Nonstructural Earthquake Damage: A Practical Guide,* published by FEMA in 1985. Available from FEMA as FEMA-74.

205. *The Homeowner's Guide to Earthquake Safety,* California Seismic Safety Commission, 1992, 1900 K Street, Suite 100, Sacramento, CA 95814-4186.

206. *Home Buyer's Guide to Earthquake Hazards,* BAREPP, 1989, 101 8th Street, Suite 152, Oakland, CA 94607.

207. *Strengthening Woodframe Houses for Earthquake Safety,* BAREPP, 1990 (ABAG P90004BAR), 101 8th Street, Suite 152, Oakland, CA 94607.

208. *Checklist of Nonstructural Earthquake Hazards,* BAREPP, 101 8th Street, Suite 152, Oakland, CA 94607.

209. *A Guide to Repairing and Strengthening Your Home before the Next Earthquake,* BAREPP, 101 8th Street, Suite 152, Oakland, CA 94607.

210. *The Home Builder's Guide for Earthquake Design 1980 Version,* Applied Technology Council, 3 Twin Dolphin Drive, Redwood City, CA 94065.

211. *Bolt-It-Down, A Homeowner's Guide to Earthquake Protection,* International Conference of Building Officials, 5360 Workman Mill Road, Whittier, CA 90601-2298.

212. *Earthquake Safeguards,* American Plywood Association, P.O. Box 11700, Tacoma, WA 98411-0700.

213. As part of the National Earthquake Hazards Reduction Program (NEHRP), FEMA has developed more than 100 publications in its Earthquake Hazards Reduction Series. A list of these publications, which are available free, may be requested by writing to FEMA, P.O. Box 70274, Washington, DC 20024.

Earthquake Preparedness in Schools and Offices

214. *Earthquakes: A Teacher's Package for K-6 Grades,* by the National Science Teachers Association, 1988, 280 pp. Available from NSTA Publications, 1742 Connecticut Avenue, NW, Washington, DC 20009; (202) 328-5800. Schools may obtain a free single copy from FEMA, Earthquake Programs, 500 C Street, SW, Washington, DC 20472.

215. *Living Safely in Your School Building,* 1986, 9 pp. Lawrence Hall of Science, University of California, Berkeley, CA 94720; (415) 642-8718.

216. *Employee Earthquake Preparedness for the Workplace and Home,* by the American Red Cross, 1988, 12 pp. Available by mail; send $1 to Red Cross Disaster Services, 1550 Sutter Street, San Francisco, CA 94109.

Making Buildings Safer from Earthquakes

217. *Peace of Mind in Earthquake Country,* by Peter Yanev, 1990, Chronicle Books, San Francisco, CA, 304 pp. ($12.95).

218. *Rapid Visual Screening of Buildings for Potential Seismic Hazards: A Handbook,* by FEMA, 1988, 185 pp. Free from FEMA; ask for FEMA-154.

219. *Reducing the Risks of Non-structural Earthquake Damage: A Practical Guide,* 1988, 86 pp. Association of Bay Area Governments (ABAG), P.O. Box 2050, Oakland, CA 94064-2050; (415) 464-7900.

220. *Strengthening Wood Frame Houses for Earthquake Safety,* 1990, 36 pp. ABAG, P.O. Box 2050, Oakland, CA 94064-2050; (415) 464-7900.

221. *Earthquake Hazards and Wood Frame Houses: What You Should Know and Can Do,* by M. Comerio and H. Levin, 1982, 46 pp. Center for Environmental Design Research, 390 Wurster Hall, University of California, Berkeley, CA 94720; (415) 642-2896.

222. *The Home Builder's Guide for Earth-quake Design,* by the Applied Technology Council, 1980, 63 pp.

Nature

223. *Seashells Common to North Carolina,* by Hugh Porter and Jim Tyler, 1971, is an excellent handbook written for the layman that describes and illustrates seashells. An essential reference for Carolina shell collectors. Available from UNC Sea Grant.

224. *The Audubon Society Field Guide to North American Seashells,* by Harold A. Rehder, 1981. This well-illustrated reference is an excellent handbook for the serious shell collector. Published by Alfred A. Knopf, New York, NY.

225. *A Field Guide to Southeastern and Caribbean Seashores: Cape Hatteras to the Gulf Coast, Florida, and the Caribbean,* by Eugene H. Kaplan, 1988. One of the Peterson Field Guide Series, sponsored by the National Audubon Society and the National Wildlife Federation. This 425-page guide describes most of the wildlife you are likely to encounter in South Carolina. Published by Houghton Mifflin, Boston, MA.

226. *The Audubon Society Field Guide to North American Fishes, Whales, and Dolphins,* by H. Boschung and others, 1983, provides detailed species accounts for the fishes and marine mammals of North America. Illustrated with color photographs. Published by

Alfred A. Knopf, New York, NY.

227. *Field Guide to Saltwater Fishes of North America,* by A. McClane, 1978, has detailed family and species accounts of common saltwater fishes. Illustrated with color drawings. Published by Holt, Rinehart and Winston, New York, NY.

228. *The Audubon Society Field Guide to North American Seashore Creatures,* by N. Meinkoth, 1981, gives detailed species accounts and an overview of taxonomy of the major shore animals of North America. Illustrated with color photographs. Published by Alfred A. Knopf, New York, NY.

Bibliographies and Newsletters

229. Publications of the Natural Hazards Research and Applications Information Center, Boulder, Colorado. Hundreds of publications are available—most for a very nominal charge—covering all aspects of natural hazard preparedness, response, mitigation, and planning. Contact the NHRAIC Publications Clerk, Campus Box 482, University of Colorado, Boulder, CO 80309-0482; (303) 492-6819.

230. *Bibliography of Publications prior to July 1983 of the Coastal Engineering Research Center and the Beach Erosion Board,* by A. Szuwalski and S. Wagner, 1984, lists coastal research published by the USACOE, available from CERC.

231. *List of Publications of the U.S. Army Engineer Waterways Experiment Station,* vol-

umes 1 and 2, by R. M. Peck, 1984 and 1985. This list updates reference 230 and lists publications by other research branches of the Waterways Experiment Station. Available from the Special Projects Branch, Technical Information Center, U.S. Army Engineer Waterways Experiment Station, P.O. Box 631, Vicksburg, MS 39180.

232. FEMA *Publications Catalog,* October 1993, is a booklet listing more than 300 publications available from FEMA designed to assist everyone from individual property owners to emergency managers. Request publication FEMA-20.

233. *Coastal Heritage,* a quarterly newsletter of the South Carolina Sea Grant Consortium, is available from the consortium, as are numerous other relevant publications. Contact South Carolina Sea Grant Consortium, 287 Meeting Street, Charleston, SC 29401.

234. The *Earthquake Education Center Newsletter* is available from the Earthquake Education Center, Charleston Southern University, P.O. Box 118087, Charleston, SC 29423-8087. The newsletter provides useful information on seismic hazards, calendars of workshops and meetings, summaries of earthquake events, and related stories. The center produced an information brochure titled *South Carolina Earthquakes.* A 10-minute video, *Earthquake Awareness and Preparedness,* is available from the center for a fee.

Federal Legislation

235. National Flood Insurance Act of 1968 (P.L. 90-448), enacted on August 1, 1968. (1) 42 U.S.C. secs. 4001 et seq. (1976). (2) 82 Stat. 476, Title 13.

236. Coastal Zone Management Act of 1972 (P.L. 92-583), enacted on October 27, 1972. (1) 16 U.S.C. secs. 1451 et seq. (1976). (2) 86 Stat. 1280. This act was amended on January 2, 1975 (P.L. 93-612; 16 U.S.C. secs. 1454, 1455, and 1464; 88 Stat. 1974), and on July 26, 1976 (Coastal Zone Management Act Amendments of 1976, P.L. 94-370; 5 U.S.C. sec. 5316, 15 U.S.C. sec. 1511a, and 16 U.S.C. secs. 1451, and 1453–64; 90 Stat. 1013).

237. Federal Water Pollution Control Act Amendments of 1972 (P.L. 92-500), enacted on October 18, 1972. (1) 33 U.S.C. secs. 1251 et seq. (1976). (2) 86 Stat. 816.

238. Marine Protection, Research and Sanctuaries Act of 1972 (P.L. 92-532), enacted on October 23, 1972. (1) 33 U.S.C. secs. 1401 et seq. (1976). (2) 86 Stat. 1052.

239. Flood Disaster Protection Act of 1973 (P.L. 92-234), enacted on December 31, 1973. (1) 42 U.S.C. secs. 4001 et seq. (1976). (2) 87 Stat. 975.

240. Water Resources Development Act of 1974 (P.L. 92-251), enacted on March 7, 1974. (1) 16 U.S.C. secs. 4601-13, 4601-14, 460ee (1976); 22 U.S.C. sec. 275a (1976); 33 U.S.C. secs. 59c-2, 59k, 579, 701b-11, 701g 701n, 701r, 701r-1, 701s, 709a, 1252a, 1293a

(1976); 42 U.S.C. secs. 1962d-5c, 1962d-15, 1962d-16, 1962d-17 (1976). (2) 88 Stat. 13, Title 1.

241. Coastal Barrier Resources Act (P.L. 97-348), enacted on October 18, 1982. (1) 16 U.S.C. secs. 3501 et seq. (2) 96 Stat. 1653.

State Legislation

242. Coastal Tidelands and Wetlands, Chapter 39 of Title 48 of the 1976 Code, as amended, by the DHEC-OCRM, 1994.

243. The Beachfront Management Act of 1988, as amended, 1990, became sections 250–360 of the Coastal Tidelands and Wetlands Act (Chapter 39 of Title 48).

Index

Stono Inlet, 107, 109, 110, 111

Storm(s), 41; damage, 71; forces, 158, 159; hazards, 29–34, 57–68, 151; protection, 5, 44; refuges, 98. *See also* Hurricane(s)

Storm surge, 11, 13, 29, 31, 34, 40, 45, 59, 68, 142, 148, 151; ebb scour, 13, 35–36, 71; flooding, 30, 40, 50, 57, 60, 63, 175

St. Phillips Island, 121, 128

Subsidence, 116, 167

Sullivans Island, 2, 4, 7, 8, 9, 10, 22, 32, 35, 61, 100, 102, 103–105, 137

Surfside, 82, 86

Surfside Beach, 84, 85

Surge, 99. *See also* Storm surge

Swash inlets, 20, 74, 78–82, 113, 130

Taking of property, 183

Terminal groin, 48, 91, 125

Thomas Yawkey State Wildlife Refuge, 95, 97

Tidal: creeks, 19, 20; deltas, 20, 46, 107. *See also* Inlets

Tides, 11, 28, 30, 148; tidal range, 19, 20, 74, 98, 116, 120

Tilghman Beach, 11, 77

Tornadoes, 36, 161

Transgressive islands, 16, 21

Trenchards Inlet, 127, 128

Tropical depression, 29

Tropical storms, 9

Truths of the Shoreline, 53

Turtle Island, 132

Undergrowth, 23, 24

Uniform Building Code, 186

Unreinforced masonry. *See* Construction

Unstabilized inlet (erosion) zone, 181

Upton-Jones Amendment, 178

U.S. Army Corps of Engineers, 28, 41, 44, 46, 87, 180

U.S. Census Bureau, 74

U.S. Environmental Protection Agency, 186

U.S. Geodetic Survey, 46

Vegetation, 13, 24, 33, 36, 41, 54, 57, 58, 60, 61, 62, 63, 65

V-zone, 71, 175

Waccamaw Neck, 86, 93

Waccamaw River, 7

Wadmalaw Island, 111

Waites Island, 21, 53, 74, 75

Walkovers, 13, 62

Wall. *See* Construction

War Between the States, 4, 8, 32, 104

Washout (Folly Beach), 37, 38, 105, 106, 107, 108, 109

Washover. *See* Overwash

Waste disposal, 66, 111, 186

Water: contamination, 65; groundwater use, 186; quality, 186; recreational use, 186; supply, 65–66

Water Classification Standards System, 186

Wave, 15, 17, 18, 19, 20, 23, 33, 45, 50, 51, 60; erosion, 57, 63, 65; forces, 146–148, 151–153; patterns, 43; reflection, 50; refraction, 42, 116

Wetlands protection, 183, 209, 235; Section 404 Wetland Certification Process, 186

White Point Swash, 78, 80

Wind, 10, 11, 12, 19, 25, 28, 29, 31, 32, 33, 34, 35, 37, 38, 50, 57, 61, 62, 65. *See also* Construction; Storm(s)

Windborne missiles, 155, 159

Windows. *See* Construction

Windy Hill Beach, 8, 77, 78

Winter storms, 37. *See also* Northeasters

Winyah Bay, 9, 10, 15, 22, 34, 86, 95, 97

Withers Swash, 81, 83

Wood-frame construction. *See* Construction

Woodstock Fault, 137, 139

Living with the South Carolina coast / Gered Lennon . . .
[et al.].
    p.   cm.—(Living with the shore)
Includes bibliographical references and index.
ISBN 0-8223-1809-1 (cloth: alk. paper). —
ISBN 0-8223-1815-6 (paper: alk. paper)
    1. Coastal zone management—South Carolina—At-
lantic Coast.
    2. Beach erosion—South Carolina—Atlantic Coast.
    3. Shore protection—South Carolina—Atlantic Coast.
    4. Coast changes—South Carolina—Atlantic Coast.
    5. Barrier islands—South Carolina—Atlantic Coast.
    6. Atlantic Coast (S.C.)—Environmental conditions. I.
Lennon, Gered. II. Series.
HT393.S6L59   1996